THE JOHN HARVARD LIBRARY

ESSAY ON CLASSIFICATION

By

LOUIS AGASSIZ

Edited by Edward Lurie

THE BELKNAP PRESS OF
HARVARD UNIVERSITY PRESS
Cambridge, Massachusetts
1962

Distributed in Great Britain by Oxford University Press, London

Library of Congress Catalog Card Number 62–19211
Printed in the United States of America

CONTENTS

EDITOR'S INTRODUCTION

In 1855, Louis Agassiz wrote a Swiss friend regarding his present condition and future prospects:

> I have now been eight years in America, have learned the advantages of my position here, and have begun undertakings which are not yet brought to a conclusion. I am also aware how wide an influence I already exert upon this land of the future, an influence which gains in extent and intensity every year.[1]

It was precisely this appreciation of his stature in American society that prompted Agassiz to plan the publication of a series of monographs for the instruction of the public on the meaning of natural history. During the next few years Agassiz worked intensely, aided by the spiritual and monetary support that had always blessed his popular efforts. In 1857 his labors were rewarded when Little, Brown, and Company published the first two volumes of his *Contributions to the Natural History of the United States.* The oversized, heavy, lavishly illustrated, finely printed volumes were in themselves physical testimony to the dominance of their author in national and international science and culture. The greater portion of volume one was entitled *Essay on Classification,* in which Agassiz delineated the theoretical principles and philosophy of his craft that made comprehensible the subsequent special studies. The *Essay* was the intellectual core of Agassiz's monumental publication, and he was very proud of it. Its pages contained the fundamental truths that had guided his long and distinguished examination of nature's creations. At a time when naturalists were debating divergent conceptions of the meaning of natural history he confidently expected his treatise to be the full and final explanation of the organic world.

In 1859, at the behest of "friends in whose opinion I have the

[1] Agassiz to Oswald Heer, January 9, 1855, Agassiz Papers, Houghton Library, Harvard University.

greatest confidence," Agassiz published a separate edition of the *Essay* in London. The volume appeared just a few months before the publication of Charles Darwin's *Origin of Species.* Those naturalists who had urged Agassiz to bring his book before the English public — Sir Richard Owen, William Buckland, and Adam Sedgwick — found Darwin's ideas intolerable. The *Essay* and the *Origin* represented two entirely opposed interpretations of nature, and no sharper contrast between the assumptions of special creationism and the concept of the evolution of species ever appeared than in the language of these two volumes. It is instructive to compare this language. Agassiz is always affirmative and assertive, his words carrying the certainty of tradition, the confidence of established truth, and the forcefulness of belief provided by a concept of ultimate causation. These tones are in sharp contrast to the patient, humble manner in which Darwin, asking for provisional intellectual acceptance, piled fact upon fact. Having staked so much on the *Essay,* Agassiz could never understand why Darwin's work was accorded such a reception; he thought he had demolished all such notions of "development" in the *Essay,* and not until 1873, the year of his death, could he bring himself to a reluctant admission that Darwin's method and proofs differed from those of earlier advocates of evolution. If for no other reason, the *Essay* is remarkable in that two years before Darwin published, Agassiz took up and answered to his satisfaction the great majority of the problems which had led Darwin to the study of the idea of evolution.

Born in Motiers, Switzerland, in 1807, Agassiz determined at an early age to become "the first naturalist of his time." Pursuing this aim with energy and ambition that never yielded to circumstance, he gained an education in natural history at the universities of Zurich, Heidelberg, and Munich. At Munich Agassiz came under the influence of the embryologist Ignatius Döllinger, the zoologist Lorenz Oken, and the philosopher Friedrich Schelling, all of whom did much to shape his thinking about the techniques and import of the study of natural history. His intellectual life in fact always reflected two important results of his German experience — dedication to exact research and a view of nature as illustrative of cosmic purpose. From Munich Agassiz journeyed to Paris, where he spent a short but significant period studying under the direction of Baron Georges Cuvier, that dominant figure in early nineteenth-century natural science whose studies of fossil remains had done much to give to paleontology its modern foundation.

Men require heroes and models, especially men with such high ambition and determination as Agassiz. For the Swiss naturalist Cuvier filled this role perfectly. When he arrived in Paris in 1832 Agassiz had one primary purpose: to gain an encyclopedic knowledge of the magnificent collection of fossil fishes housed in the museum of paleontology at the Jardin des Plantes. Such competency would provide basic data in an uncharted field and add significantly to the foundations of paleontology. The rich Paris materials were under the supervision of Cuvier, whose immediate high regard for Agassiz led him to place all such data completely at his disposal. Cuvier, moreover, convinced that Agassiz was doing work of fundamental value, turned over to him his own drawings and research notes on fossil ichthyology gathered in the course of years of study. Cuvier's death a few months after Agassiz came to Paris profoundly affected the young Swiss — Agassiz would always think of himself as having inherited both the technical competence and the philosophy of nature that distinguished Cuvier's work.

In a paper of 1812 and in the two editions of *Le Règne animal* (1817 and 1829–30) Cuvier set forth a system of classification based on the concept of four distinct branches in the animal kingdom, each typified by a different anatomical plan of structure. This morphological approach to taxonomy stressed the anatomical identity of forms within particular branches — mollusks, radiates, articulates, and vertebrates — but denied that these branches shared any genetic relationship with each other. Each branch was present in nature from the beginning, its character and permanence the result of supernatural inspiration. The branches contained units of taxonomic identity in an ascending scale of inclusiveness — species, genera, families, orders, and classes. The finite individuals of animate nature were forever linked to the immaterial forms in which they participated, as represented by the divinely inspired identity of structural plan. This view of nature, which reached back to Plato for its basis, influenced all of Agassiz's experience with the facts of organic creation. He became, after Cuvier, the most steadfast and articulate exponent of the idea of divine intervention in natural history. His admiration for Cuvier's exposition of this philosophy was such that in the *Essay on Classification,* the work he considered his supreme intellectual achievement, he called the French savant "the greatest naturalist of all times."

It was in Paris, too, that Agassiz enjoyed the friendship and support of the Prussian statesman, philosopher, and naturalist, Baron Alexander von Humboldt. While Cuvier taught Agassiz much about the method and philosophy of science, Humboldt taught him less tangible but equally valuable lessons. He impressed on the young Swiss the importance of cultural and political support for scientific endeavor, introduced him into the salons of wealthy and influential patrons of science, and showed him by precept and example that the man of science must also be a man of the world. To Humboldt Agassiz owed his appointment as professor at the Collège de Neuchâtel in Switzerland, and to him the younger man could always turn for spiritual and material aid. Indeed, the special virtues of both Humboldt and Cuvier helped produce in Agassiz a brilliant investigator, an inspired teacher, and a vibrant personality.

From the time he came to teach at Neuchâtel in 1832 until he left the Swiss town for the United States in 1846, Agassiz established himself as a professional naturalist of high talent and promise. In a pioneer effort he published *Recherches sur les poissons fossiles,* (1833–1843), a description and analysis of over 1,700 species of ancient fishes, detailing their anatomy, geographical distribution, zoological character, and stratigraphic relationships. This work, an amazing feat for so young a man, earned him European recognition. His analyses of fossil fishes were models of the kind of precise investigative techniques he had learned from Döllinger and Cuvier. In these five volumes his fellow naturalists could see spread before them the rich holdings of the great and small European museums and of private collections as well.

Even at this early stage in his career Agassiz showed a strong interest in classification. Believing that "a physical fact is as sacred as a moral principle," he understood that facts by themselves were of little value unless interpreted. He devised a new system of classification for fossil fishes, drawn from principles of comparative anatomy. In *Poissons fossiles* he gave evidence of discarding one of the major influences of his German education, the concept of *Naturphiloso-phie,* that romantic view of organic creation fathered by his teachers, Oken and Schelling. Agassiz suspected the "speculative" tendencies of the doctrine as too threatening to the concept of permanence in the universe, since it suggested that species bore relationship to one another in some unified order of derivation and development. Instead, study of ancient fishes convinced him that these were "phenomena closely allied in the order of their succession, and yet without sufficient cause in themselves for their appearance." The character of fossil fishes was the result of a "primary plan" instituted by a "superior intelligence whose power . . . established such an order of things."

In still another field Agassiz was able to supply empirical demonstrations for metaphysical assumptions. His notable paper before a local Swiss scientific society announced in 1837 his agreement with the findings of earlier investigators that the peculiar configurations of the land and the placement of boulders and other debris had been the result of ancient glacial action. From this theory Agassiz reasoned that much of northern Europe had known a vast Ice Age in the period of its history immediately prior to the modern epoch. (It is now more accurate to speak of "ice ages," since subsequent study showed that perhaps as many as four or five advances and retreats of ice had occurred with intervening warm interglacial periods.) The glaciers of modern times were contemporary evidence of what had taken place in the recent past. The establishment of the "Eiszeit" concept as the primary force in the natural history of the Pleistocene epoch was another singular landmark in Agassiz's march along the road to becoming the first naturalist of his time. He substantiated his initial analysis by two books published in 1840 and 1847,[2] the result of much painstaking investigation in the Alpine regions of his native land and in England, Scotland, and Germany.

Agassiz's own research in geology would furnish evolutionists with excellent data with which to refute doctrines of special creationism. In the view of men of a different persuasion the advance and retreat of glacial ice in northern regions was shown as a profound influence upon the geographical distribution and zoological character of contemporary flora and fauna. The movement southward of glacial ice

[2] *Études sur les glaciers* (Neuchâtel, 1840), and *Système glaciaire* (Paris, 1847).

in parts of Europe and North America resulted in the successive transformation of once tropical regions to temperate zones, pine forests, barren tundras, and slowly moving masses of ice. As the ice advanced, some arctic forms moved southward and established themselves in tropical regions. Some animals returned to their former homes with the retreat of the ice and the revival of warmth, but others remained permanently in their new locales. Animals and plants in the path of glacial advance either moved further south, were extinguished by the cold, or adapted to the new conditions of life. These changes in zoological geography resulted in the contemporary admixture of arctic forms with those of temperate and tropical regions. Glaciation and geographic changes related to it thus explained the dispersal of flora and fauna from their original habitat, and the consequent operations of natural selection in preserving successful variations that suited new conditions of life.

Agassiz, however, found it impossible to interpret the results of his research in such a fashion as to provide a positive role for "physical agencies" in natural history. Once again, offering substantiation for Cuvier's interpretations, he saw the glaciers as "God's great plough," destructive forces that yet signified supernatural intervention. To Agassiz the glacial period was a magnificent albeit chilly demonstration of the power of the Deity to cause great catastrophes, events which had wiped out life in previous epochs, after which it was created anew by divine action. Therefore, it was impossible for any genetic relationship to exist between past and present forms, so firm was his opposition to concepts of unity of origin and common development in nature. As Agassiz characterized recent natural history,

> There is . . . a complete break between the present creation and those [creations] which precede it; if the living species of our times resemble those buried in the levels of the earth, so as to be mistaken for them, it cannot be said that they have descended in direct line of progeniture, or what is the same thing, that they are identical species.[3]

The early achievements of his career established Agassiz as a wide-ranging scientist, capable of generalizing about nature as a result of distinguished special studies in geology, paleontology, and ichthy-

[3] Quoted in Jules Marcou, *Life, Letters, and Works of Louis Agassiz* (2 vols., New York, 1896), I, 107.

ology. Moreover, he made significant contributions to natural history by publishing French and German translations of important English monographs, compiling a comprehensive bibliography of zoology, and writing a valuable treatise on zoological nomenclature. The format of these accomplishments was no less distinguished than their content; Agassiz established his own publishing and engraving house at Neuchâtel and produced works notable for their design and illustrations.

These years of substantial intellectual attainment and success earned Agassiz a reputation that spread all over Europe and reached America as well. Men of such renown as Darwin, Sir Charles Lyell, William Buckland, Adam Sedgwick, Leopold von Buch, and Élie de Beaumont ranked him as their peer and delighted in the productions of his pen. Thoroughly familiar with the treasures of the great museums of Europe, honored by membership in the outstanding learned societies of England and the continent, awarded prizes and research grants by the governments and scholarly institutions of Prussia, France, and England, Agassiz enjoyed the rank of a leading naturalist in his generation. His fortunate combination of personal ambition, energy, and talent enabled him to equate private advancement with public progress. Never at a loss for grand projects planned with high enthusiasm, Agassiz thoroughly enjoyed his work because in it he fulfilled himself. Fundamental to his intellectual character was the conviction that special knowledge to be truly valuable had to be shared with the intelligent populace. This transmutation of subjective motivation to objective purpose became a basic force in his life, that of a conspicuous public man whose zeal for natural history infected all who heard or read him.

Such personal and intellectual attributes particularly impressed and appealed to Americans in 1846, the year Agassiz arrived in Boston to lecture at the Lowell Institute and to begin his two-year research period in the United States under the auspices of the Prussian monarchy. Hailed as a "capital fellow" by scientists proud to have landed such a "big fish" from Europe, and greeted with universal acclaim, Agassiz basked in the glory and power of a great public reputation. His remarkable ability to translate the facts of nature into an idiom at once understandable and inspiring, thrilled all classes of Americans. His new admirers marveled at their good fortune in having such a personage in their midst and were overjoyed when

Agassiz decided to make America his permanent home. Comfortably established as professor of zoology and geology at the Lawrence Scientific School of Harvard University by 1847, Agassiz welcomed in equal measure the adulation of Ohio fishermen and the good will of poets like Longfellow and Lowell. His dedication to national cultural progress was time and again evident as he spurned offers to return to the continent of his birth. He repaid his admiring New World public by extolling the grandeur of the natural environment and boasting of America's ability to equal and even surpass the intellectual achievements of Europe. Agassiz lent his own prestige to emphasis on the need for professional standards in science and the importance of public support for scholarly endeavor. He played a major role in the organization of the American Association for the Advancement of Science, acted to improve the character of higher education in the sciences, and strove to create new institutions for research and instruction. One notable achievement was the establishment of the Museum of Comparative Zoology at Harvard College in 1859, the direct result of years of public pleading by Agassiz and of his magical ability to raise money for science from private benefactors and governmental bodies. Such involvement in the social relations of science made Agassiz a commanding figure in American culture. He was ambitious, energetic, and dedicated; his private goals were public aims, and his drives reflected the ambitions of America in the exciting religion of cultural progress that characterized the decades before the Civil War. Understandably, Emerson was one of Agassiz's greatest admirers, applauding the nobility of his cultural purposes, the virtue of his social democracy, and the validity of his natural religion.

The planning and execution of the *Contributions to the Natural History of the United States* and its prefatory *Essay on Classification* were primary examples of the advantages and the penalties of Agassiz's role in American society. From his first days in the New World he had been fascinated with the rich, and as yet barely described, natural history of America. Between 1846 and 1854 he had undertaken a number of exploratory journeys through the eastern seaboard states, the deep South, to the White Mountains, to the Lake Superior region, and up the Mississippi valley. In these he had garnered a tremendous array of materials of every kind, so much in fact, that Harvard University had to make two buildings available to him for storage alone. Even these were not sufficient, and by 1859 Agassiz's insatiable appetite for collecting had made necessary the establishment of the Museum of Comparative Zoology. He saw his collections as the primary materials with which to educate American students of natural history according to the high standards of European scholarship, but as his commitment to education grew and his public activity increased, these involvements had serious intellectual consequences. Some colleagues, while applauding his zeal for collecting, thought such efforts detracted from the primary task of studying the specimens themselves. Although Agassiz had published a few papers describing his American researches, up to 1855 he had done nothing in the United States to compare with the work of his European years. After the lapse of nearly a decade, he had published but two books, one a textbook of joint authorship and the other an account of his exploration of Lake Superior.

Agassiz himself began to see that his labors of recent years had interfered with intellectual effort and determined to reassert his scholarly pre-eminence. As he wrote Sir Charles Lyell in 1856, he planned to devote himself henceforward to scholarly publication, and his forthcoming *Contributions* would show the world "that I have not been idle during ten years' silence." As with so many projects of the past and the future, his plan for a major publication had begun on a small scale and grew larger from the stimulus of his imagination. It became a grandiose scheme involving many supporters and considerable financial assistance.

Returning from a journey to the lower South and the Mississippi valley in 1853, Agassiz was thrilled as always by first-hand examination of nature. In this case it was the fishes typical of the Tennessee and Mississippi river systems that fascinated him, and he planned to publish a comprehensive *Natural History of the Fishes of the United States.* After enlisting the cooperation of scores of amateur and professional collectors and of the federal government as well, Agassiz had accumulated ample materials for this purpose. Instead of going ahead with his task, however, the success of his collecting enterprise impelled him to embark on an even grander quest for still more specimens representative of all American natural history and to merge this data with the extensive collections he had in Cambridge. The result of his labors would be a series of ten monographs that would detail the entire natural history of the United States.

Thus the idea of the *Contributions to the Natural History of the United States* was born in the imagination of the man who considered himself America's first naturalist. It was appropriate that he turned to the populace at large for material support, since his books would represent

> an American contribution to science, fostered and supported by the patronage of the community. . . . I hope in this way to show my friends in Europe that American naturalists have entered upon a fair competition with the scientific labors of the old world, and that they aspire, with a generous ambition, to achieve their scientific independence . . .[4]

In this venture the congenial alliance Agassiz had made with rich and influential men in New England served him well. Francis Calley Gray, patron of science and wealthy manufacturer, proposed a scheme whereby the publishing house of Little, Brown, and Company agreed to produce the volumes if an advance sale of 450 subscriptions could be secured. A prospectus detailing the character of this effort undertaken, Agassiz asserted, "for no other purpose than to contribute my share towards increasing the love of nature among us," was sent to thousands of people in America and Europe.

Between 1855 and 1857 Agassiz's campaign for subscriptions was paralleled by a similar quest for still more specimens of natural history, in this case turtles and turtle eggs needed to supply zoological and embryological material for the first two volumes. It is impossible to imagine how any literate American of these years could have been unaware of Agassiz's grand effort, so widespread were the publicity and collecting activities of the naturalist and his friends. The results were predictable, since they were a simple reflection of the prestige and influence Agassiz enjoyed. By late 1856 he had thousands of turtles, was assured of over 2,500 subscribers and at least $300,000 in advance sales for the entire printed series. "I do not think Humboldt himself could obtain in all Europe . . . such a subscription for so expensive a work," Agassiz proudly reported to

[4] Printed *Prospectus* announcing *Contributions to the Natural History of the United States,* May 28, 1855, p. 4, Agassiz Papers, Houghton Library, Harvard University.

his friend Charles Sumner. Subscribers were promised books that would illuminate "the wonderful diversity of the animal creation of this continent" by describing all classes of animals, such natural history to be illustrated by accurate and expensive drawings. The first monograph would portray the embryology, geographical distribution, and classification of North American turtles. Most important, these special studies would be preceded by an "Essay on Classification" that would supply "a new foundation for a better appreciation of the true affinities, and a more natural classification, of animals." This venture in taxonomy, Agassiz promised, would enable naturalists to determine "with considerable precision, the relative rank of all the orders of every class of animals" and would furnish "a more reliable standard of comparison between the extinct types of past geological ages and the animals now living upon the earth." [5]

When the first two volumes comprising the *Essay* and the analysis of turtles were published in the fall of 1857, Agassiz was very proud. He was acutely aware of their importance. His words to a colleague when just beginning the project were revealing: "I have tried to make the most of the opportunities this continent has afforded me. Now I shall be on trial for the manner in which I have availed myself of them." [6] But it is doubtful whether the admiring public learned very much from the intricate and highly specialized descriptions of North American *Testudinata*. They could not fail to be instructed by the *Essay on Classification*.

Agassiz asserted in his preface that the *Essay* would be of interest and value to all. It would serve as a "text-book of reference for the student, in which he may find notices of all that has been accomplished in the various departments of Natural History." Moreover, he boasted that his audience was so wide that "I expect to see my book read by operatives, by fishermen, by farmers, quite as extensively as by the students in our colleges or by the learned professions." The *Essay* would call attention to "the mode of life of all our animals, to their geographical distribution, their natural affinities, their internal structure, their embryonic growth, and to the study

[5] *Ibid.*, pp. 3–4.
[6] Agassiz to Samuel S. Haldeman, May 31, 1855, Samuel S. Haldeman Papers, Academy of Natural Sciences of Philadelphia.

of fossil remains."[7] Americans could hardly ask more of their leading interpreter of nature.

In the *Essay* Agassiz attempted to write a modern version of Cuvier's taxonomy. He set himself the task of determining whether systems of classification from Aristotle forward to the 1850's were "true to nature." The result was almost foreordained. After appraising many prior efforts at classification, Agassiz announced:

> I am daily more satisfied that the primary divisions of Cuvier are true to nature, and that never did a naturalist exhibit a clearer and deeper insight into the most general relations of animals than Cuvier. . . . (pp. 146–147)

But the *Essay* was not a work reflecting the simple adulation of a student for a teacher. Agassiz set out to demonstrate that the concept of classification central to the philosophy of special creationism was objectively valid and not merely a product of subjective human invention. Cuvier was elevated to the highest rank among taxonomists, not because he was Cuvier, but because his system reflected the reality of nature itself.

Agassiz began this demonstration by posing a question as fundamental for modern zoology as for his own time. Was zoological classification an artificial effort by the naturalist to impose his own subjective values upon nature? Or was there, instead, an objective basis for the work of the taxonomist that would give his findings the validity of natural law? As he phrased the problem:

> Are these divisions artificial or natural? Are they the devices of the human mind to classify and arrange our knowledge . . . or have they been instituted by the Divine Intelligence as the categories of his mode of thinking? (p. 8)

In this contrast of values idealism was classed as objective and the finite operations of the human mind as mere approximations of truth. The task of the modern naturalist, therefore, was to contrive to reflect in his descriptions and classifications the reality of the natural world that comprised the operations of the Deity. He did this by understanding the fundamental rationality of a divinely ordained world, planned in perfect fashion from the beginning, always exemplifying the wisdom of the creative intention. Understanding came through the intensive application of intellect; such rationality

[7] *Contributions to the Natural History of the United States* (4 vols., Boston, 1857–1862), I, viii–x.

gave the naturalist an awareness of ultimate truth. For Agassiz, therefore, there was

> a system in nature to which the different systems of authors are successive approximations. . . . This growing coincidence between our systems and that of nature shows . . . the identity of the operations of the human and the Divine intellect; especially when it is remembered to what an extraordinary degree many *à priori* conceptions . . . have in the end been proved to agree with the reality, in spite of every objection at first offered by empiric observers. (pp. 25–26)

Agassiz in effect equated the work of the naturalist with that of a theologian. The taxonomist's understanding of Divine intent was made precise by empirical knowledge of nature so that he could report on the conditions of creation with the greatest accuracy. Of the men who strove to achieve this correspondence between subjective perception and divine reality Cuvier was by far the most significant.

In so defining the nature and purpose of classification and its practitioners, Agassiz asked a modern and pertinent question and supplied a traditional answer. His view that classification should reflect the actuality of nature was advanced, but his conception of what that reality was stretched back to Plato, the Greek speaking through the more recent perceptions of Cuvier. Agassiz, in fact, broadened the scope of his master's idealism by making it more inclusive. Affirming with Cuvier that the identity of structure of the major branches of the animal kingdom signified an "intellectual conception which unites them in the creative thought," Agassiz ascribed the same ideal character to all the lesser taxonomic divisions. Thus, "the species is an ideal entity, as much as the genus, the family, the order, the class. . . ." As an ideal form, the species continued to exist; even though the individual animal that represented it on earth perished, it was replaced by others that signified the same ideal type. Individuals, too, exemplified in their form and structure the entire range of higher, ideal categories and the naturalist could discern such features in them.

> Individuals . . . represent their species . . . [and] at the same time their genus, their family, their order, their class, and their type [branch], the characters of which they bear as indelibly as those of the species." (p. 176)

Agassiz always insisted that his metaphysical conclusions were grounded upon the actuality of nature itself. This intellectual trait identified him with both past and present, and made him at once so perplexing and so admired an intellect to men like Darwin and Lyell. The only valid complaint such colleagues could make of Agassiz was that he had become a victim of the concepts and heroes of his youth. On this fact rests one of the primary reasons why the *Essay on Classification* is such a notable document in the history of ideas. It charts the course of scientific and philosophic education within a tradition that gave singular emphasis to man's special conception of the cosmos — that of the supremacy of immaterial purpose in the universe. At the same time it demonstrates the intellectual power derived from the exact investigation of natural facts. As a result, Agassiz is able to offer some highly pertinent and modern appraisals of problems in zoology, as well as to point up deficiencies in the idea of development. Finally, the *Essay* shows how the metaphysical interpretation of the facts can lead only to idealistic conclusions for an Agassiz, whereas the same facts meant something entirely different to a man of Darwin's outlook.

In the 1840's most professional naturalists were opposed to the concept of development. This was true not only of Agassiz, but also of such future Darwinians as Thomas Henry Huxley and Agassiz's Harvard colleague, the botanist Asa Gray. Gray had applauded the refutation of materialism continued in Agassiz's first Boston lectures of 1846–47, in which the Swiss naturalist had ridiculed the notions of unity and common serial development in the animal kingdom advanced by the then unknown Robert Chambers in *Vestiges of the Natural History of Creation* (1844). This was an easy task because such arguments were highly speculative, being unsupported by much direct evidence. In the *Essay,* similarly, it took little effort for Agassiz to refute the assertions of crude environmentalists like Lamarck. He merely marshaled the preponderance of known facts against vague assertions of physical unity in animals and of the inheritance of acquired characteristics. It was a primary characteristic of the intellectual divergence between advocates of special creationism like Cuvier and Agassiz and proponents of the developmental hypothesis such as Chambers, Lamarck, or Étienne Geoffroy Saint-Hilaire, that upholders of classical biology were far superior in their actual experience of nature and their command of specialized subjects.

In contrast to the earlier period, the decade of the 1850's, when

Agassiz planned and wrote the *Essay,* was a time of significant conceptual transformation in biology. These were the years when Darwin, Huxley, Joseph Dalton Hooker, and Gray worked at the elaboration of a new conceptual framework for natural history, one that would supplant what they had come to regard as the sterile and outmoded concept of special creationism. Gray's and Hooker's researches had shown, for example, that there were important resemblances and affinities between plants of one geographic region and those of another. Such findings made it more reasonable to suppose that animal and plant forms had originated from single pairs in common centers of creation. Their present dispersal and divergencies from the original stock could be accounted for by the action of physical agencies such as the effects of climate in the present or the action of glaciers in the past. While Darwin would elaborate such findings into a general theory of variation as the result of natural selection, one did not have to be a convinced evolutionist to believe in the unity of origin of zoological forms.

Naturalists identified with this new intellectual persuasion were notable for one common trait: they wished for a free interchange of theory and opinion regarding the new data that were being discovered. Thus Gray wrote the Yale naturalist, James Dwight Dana, in 1856

> The right way to bring a series of pretty interesting general questions towards settlement is perhaps in hand . . . *viz.,* for a number of totally independent naturalists, of widely different pursuits and antecedents—to environ it on all sides, work towards a common centre, but each to work perfectly independently. Such men as Darwin, Dr. Hooker, [Alphonse] De Candolle, Agassiz, and myself, — most of them with no theory they are bound to support, ought only to bring out some good results.[8]

As Gray would discover, Agassiz was wedded to the defense of special creationism. This unyielding conviction set him apart from some fellow naturalists in the years immediately preceding and subsequent to the publication of Darwin's work. By the 1860's Agassiz's attitude made him one of the few scientists in America publicly opposed to a conception of nature like evolution not grounded on a belief in the permanence of species imposed from without.

For example, Agassiz denied the possibility that animals and

[8] December 13, 1856, in Jane Loring Gray (ed.), *Letters of Asa Gray* (2 vols., Boston, 1893), II, 424.

plants had originated in common centers of creation, their present distribution being the result of such physical factors as glacial action. The Ice Age had provided, instead, a physical barrier to any unity of development between past and present. As he wrote in the *Essay*,

> It appears to me . . . as facts now point distinctly to an independent origin of individuals of the same species in remote regions, or of closely allied species representing one another in distant parts of the world, one of the strongest arguments in favor of the supposition, that physical agents may have had a controlling influence in changing the character of the organic world, is gone for ever. (p. 39)

How, then, explain the particular patterns of geographical distribution? "Only the deliberate intervention of an Intellect, acting continuously according to one plan, can account for phenomena of this kind." From these statements it seemed to some that Agassiz had become more dogmatic with the years and had remained uninfluenced by the course of zoological research. To Lyell, therefore, the *Essay* showed that Agassiz had "nailed his colors to the mast" and announced for all time his opposition to ideas of change and development.

Yet when naturalists like these men criticized Agassiz for his reliance on old standards, they were in a sense castigating their own high admiration for the idealism that had seemed attractive to Gray when he first heard Agassiz in 1846. Agassiz had not changed; his audience had. Moreover, while it was true that his social involvements had not allowed him the time or interest to become familiar with new directions in research, active investigators — Karl Ernst von Baer and Sir Richard Owen for example — were nevertheless equally firm opponents of the developmental concept. Two years before Darwin, it was still true according to all the standards of science Agassiz and men like him respected, that the idea of development was still an unproved hypothesis. He could hardly have been expected to support the theory before 1859, when even Huxley and Lyell, unburdened by Agassiz's idealism, had not yet publicly embraced the idea, although they encouraged Darwin in his efforts. Proponents of evolution criticized Agassiz for his continued opposition in the 1860's, yet his attitude was consistent with views he had held since the 1830's. Such supporters of Darwin as Lyell and Gray were men of equal good will, but in their dissatisfaction with Agassiz—and especially with his popular attacks on evolution — they seemed to expect that he would somehow take a position quite contrary to his entire philosophical background and public role since the beginning of his career. Even in their case, such were the shadings of opinion surrounding the theory of evolution that, while giving private and public encouragement to Darwin, these two men did not commit themselves so forthrightly to the concept as Darwin later wished they had. Indeed, Agassiz's ability to search out weaknesses in the theory forced Darwinians to seek further philosophical and scientific substantiation for the concept. Although Darwin had placed the idea of development on a new plane of objectivity, he left unresolved certain fundamental problems in the evolutionary synthesis — the nature of heredity, the mechanics of variation, and the manner in which evolution operated as a phenomenon affecting entire populations. Yet, given the state of scientific knowledge in 1857, if Darwin was justified in expecting ultimate validation of his beliefs, Agassiz, in the light of his own experience and education, was equally justified in a firm reliance on his principles.

As expressed in the *Essay,* these principles comprised the most affirmative and articulate exposition of classical biology published in the nineteenth century. Constant reliance upon the assumptions of special creationism in every aspect examined resulted in Agassiz's view that "the combined intellectual efforts of hundreds of investigators" proved "the intervention of a Supreme Intellect" in the "combinations of nature." Significantly enough, in all his surveys of prior taxonomical and zoological research, Agassiz fails to give adequate representation to men of a different outlook. Thus, the work of Hooker on the geographical distribution of plants goes unnoticed, he refers to Darwin solely for a paper on marine biology while the *Zoology of the Voyage of the Beagle* (1840) is ignored, Lyell's pioneering *Principles of Geology* (1830–33) receives mention only as a text book, and Huxley is cited primarily to correct his erroneous interpretation of the work of von Baer. On the other hand, Sir Richard Owen, who pronounced the *Essay* "the most important contribution to the right progress of zoological science in all parts of the world where progress permits its cultivation," receives prominent attention.

Just as Agassiz employs bibliography to buttress special creationism, so his repetition of concepts and viewpoints is constant, leaving no room for doubt that his is the only valid approach to natural history. The *Essay* proceeds from demonstration to demonstration; after each part or discussion he provides a detailed recapitulation of what has been said before. In addition, Agassiz supplies a summary of his reasoning and beliefs in Section XXXII of the first chapter, where he presents to the reader a kind of unified credo, thirty-one "conclusions" representing the scientific and philosophical doctrines of the entire work. To Agassiz the final meaning of all natural history is that it

> exhibits not only thought, it shows also premeditation, power, wisdom, greatness, prescience, omniscience, providence. . . . all these facts . . . proclaim aloud the One God, whom man may know, adore, and love; and Natural History must, in good time, become the analysis of the thoughts of the Creator of the Universe, as manifested in the animal and vegetable kingdoms, as well as in the inorganic world. (p. 137)

Such certainty of belief held by him and like-minded naturalists was derived from a logic and method of science as persuasive as it was comprehensive.[9] Of primary significance was the conception of a rational plan in nature, imposed from without, and controlling the existence and relationships of all organized beings. In all the diversity and multiplicity of life on earth, an essential and orderly unity prevailed. The aim of the naturalist must therefore be to describe this unity by virtue of his perception of the divine in nature. In Agassiz's language there were

> fixed relations between animals, determined by thoughtful considerations. I would as soon cease to believe in the existence of one God because men worship Him in so many different ways . . . as to distrust the evidence of my own senses respecting the existence of a preëstablished and duly considered system in nature, the arrangement of which preceded the creation of all things that exist. (p. 155)

The purpose of systems of classification, then, was to demonstrate "the manifold ties which link together all animals and plants as the living expression of a gigantic conception. . . ."

A traditional and ultimate interpretation of nature was not in

[9] My discussion of Agassiz's philosophy of nature is greatly indebted to the incisive appraisal by Ernst Mayr, "Agassiz, Darwin, and Evolution," *Harvard Library Bulletin,* XIII (1959), 165–194.

itself sufficient logical ground on which to deny the possibility of change in the organic world. It was possible to argue, as did supporters of the concept of design, that the Deity had created a world whose creatures advanced from simple to complex as the result of intelligent, initial, divine planning. To Agassiz and other spokesmen for classical biology this was a poverty-stricken view of the creative power. The force responsible for the plan of creation acted as consistently in the present as in the beginning. Power of this sort could not be equated with the agencies it had called into being. It could not be made synonymous with physical forces, themselves the product of immaterial design, acting to "change" the creatures of the organic world that were also examples of divine action. Flora and fauna could not "cause" their own alteration or fate in life. Rather, they existed as representations of creative design and rational categories of thought. It was impossible to conceive of these ideal types as transitory or mutable, since divine thought was permanent

> a power capable of controlling all . . . external influences, as well as regulating the course of life of every being, and establishing it upon such an immutable foundation . . . that the uninterrupted action of . . . [physical] agents does not interfere with the regular order of . . . natural existence. . . . (pp. 90–91)

Since each animal or plant signified a category of thought it was logically impossible for such types to give rise to "varieties" or to change into something else. For Agassiz here was a profound reason to contradict any concept of evolution based on the action of observable secondary phenomena:

> Nothing seems to me to prove . . . more fully the action of a reflective mind . . . than the different categories upon which species, genera, families, orders, classes, and branches are founded in nature and manifested in material reality in a succession of individuals, the life of which is limited . . . to comparatively very short periods. (p. 25)

Thus the taxonomist had to identify the particular role and rank in the everyday world of the multitude of animals and plants whose place and definition had been set forth according to a rational plan in the beginning. Agassiz, therefore, took great pains to establish a system of classification that was "true to nature," namely, one that reflected with greatest accuracy the immaterial plan of the Creator. Because specific designation and identity had to be attributed to animals and plants, he quite properly praised the contributions of

Aristotle and Linnaeus to taxonomy. In the modern period Agassiz applauded the labors of taxonomists like Cuvier and von Baer, who understood that "there is no such uniformity or regular serial gradation among animals." Since categories of thought were mutually exclusive with no genetic connection possible among the four great branches of the animal kingdom, it followed that the "plan of structure" of each was "so peculiar that we nowhere find analogies . . . extending from one branch to all the representatives of another branch." No system of classification that suggested such relationship could be "true to nature."

This discontinuous view of creation underscored Agassiz's entire conception of natural history and philosophy. Within the framework of its logic it was possible to ascribe tremendous if capricious power to the Creator, who was given responsibility for the origin of life and for all subsequent alterations. In one example after another Agassiz used this logic to refute notions of physical unity and genetic affiliation. Animals and plants of ancient times and of the present were viewed as the result of independent creative acts, their life history always subject to divine intervention in the form of catastrophes. Animals that remained unchanged after long periods of time demonstrated the wisdom of the plan; those that were extinct bore "prophetic" relationship to their successors that proved creative intent for the future.

The evidence of creative force on earth was the individual animal or plant. Each individual exemplified a species, a genus, a family, and other higher and more inclusive orders of taxonomic identity. Ideal categories never changed, so that it was impossible to speak, as evolutionists did, of the "transmutation of species." Whereas evolutionists thought of species as subjective creations of the naturalist, Agassiz insisted on their ideal character, thus also denying their biological reality but doing so from ultimate considerations. He was therefore able to ask the perfectly logical question: "if species do not exist at all, as the supporters of the transmutation theory maintain, how can they vary? And if individuals alone exist, how can the differences which may be observed among them prove the variability of species?" [10]

Agassiz was too careful an investigator to deny the fact of change

[10] *American Journal of Science,* XXX (2d ser., 1860), 143.

in nature. He never questioned variation itself, but rather the significance that advocates of development attributed to it. Equating change in nature with change in individuals, he affirmed that such individual variation never influenced the permanence of the type. Thus he could write, "I have seen hundreds of specimens of . . . our Chelonians, among which there were not two identical. . . . truly, the limits of this variability constitutes one of the most important characters of many species." (p. 66) Yet he could assert with equal conviction that "it was a great step in the progress of science when it was ascertained that species have fixed characters and that they do not change in the course of time." (p. 58)

Denying the significance of individual variation, Agassiz nevertheless attributed much importance to the development of the individual. Embryology, therefore, furnished "the most trustworthy standard to determine the relative rank among animals." Emphasis upon ontogenetic transformation led Agassiz to admire the work of Karl Ernst von Baer, the modern founder of embryological investigation. The labors of von Baer impressed him as proving that while individuals experienced a series of transformations, they "never produce anything different from the parents . . . all reach through a succession of unvarying changes, the same final result." Development was therefore a phenomenon identified solely with the life history of the individual, and such finite events could not effect the type, which was fixed and final. All so-called "change" was simply a "cycle of growth." Life cycles "revolve forever upon themselves, returning at appointed intervals to the same starting-point and repeating through a succession of phases the same course." In so defining "evolution" Agassiz also adopted another, equally inclusive viewpoint toward change in nature. He persisted in equating change in individuals or ontogeny, with phylogeny, or the history of the entire race or group. Thus the phases of development in living animals were considered recapitulations of the order of succession of their extinct ancestral forms. The history of the type was therefore the "cause" of the history of the individual, and such recapitulation of phylogeny in ontogeny demonstrated that

the leading thought which runs through the succession of all organized beings in past ages is manifested again in new combinations in the phases of the development of the living representatives of these different types. It exhibits everywhere the working of the same creative Mind, through all times, and upon the whole surface of the globe. (p. 115)

The history of the particular type-plan to which an individual belonged reflected the development of the individual, all such change representing fixed and final products. These embryological concepts, predicated on assumptions of final and absolute fixity of types, comprised the only kind of "evolution" that Agassiz could ever admit. Darwin and his followers subsequently attempted to use the concept of recapitulation as support for their doctrines. But as Agassiz and von Baer phrased and understood the notion, it was in fact a conception generally valueless for an understanding of the dynamics of change in populations.[11]

The persistence of ideal assumptions in the *Essay* should not obscure the fact that it contained some highly valuable statements and guideposts for the study of natural history. For example, the reader is told that "every new fact relating to the geographical distribution of well-known species is as important to science as the discovery of a new species." In this connection, Agassiz urged new directions for zoological research:

without a thorough knowledge of the habits of animals, it will never be possible to ascertain . . . the true limits of all those species which descriptive zoologists have . . . admitted . . . into their works. . . . what does it matter to science, that thousands of species . . . should be described and entered in our systems, if we know nothing about them? (p. 66)

In emphasizing the need for research in ecology Agassiz offered important advice to his fellow naturalists:

Every species is described as if it stood alone in the world. . . . Yet, how interesting would be a comparative study of the mode of life of closely allied species. . . . We scarcely possess the most elementary information . . . to discuss . . . the question of the instincts, and in general the faculties of animals, and to compare them together and with those of man. . . . Who . . . could believe for a moment . . . that the habits of animals are . . . determined by the circumstances under which they live, after having seen a little turtle . . . still enclosed in its egg-shell, which it hardly fills half-way, with a yolk bag as large as itself . . . with its eyes shut, snapping as fiercely as if it could bite without killing itself? (pp. 67–68)

Again, in words striking in their modernity, Agassiz states:

[11] See Agassiz's discussion of recapitulation in Chapter II, section 8, and Chapter III, section 6 of the *Essay*, and especially the notes and references in these sections.

> Is there an investigator, who having once recognized . . . a similarity between certain faculties of Man and those of the higher animals, can feel prepared . . . to trace the limit where this community of nature ceases? . . . I confess I could not say in what the mental faculties of a child differ from those of a young Chimpanzee. (p. 68)

Such statements were representative of many perceptive observations by Agassiz. In the years that followed the publication of the *Essay* naturalists who were not burdened by the limitations of special creationism contributed to the solution of problems and questions posed there.

The public and professional response to the *Essay* was a preview in microcosm of Agassiz's social and intellectual career after 1857. Humboldt's words of admiration, published widely in American newspapers, reflected Agassiz's conception of himself as a universal intellect, quite as capable of cosmic generalizations as his life-long mentor. Sir Richard Owen's appreciation typified Agassiz's dominant role as an opponent of Darwin in the scientific controversy over evolution of the early 1860's. Oliver Wendell Holmes's glowing praise for "Agassiz's Natural History" in the *Atlantic Monthly*[12] was a prediction of Agassiz's subsequent relationship to that controversy. Unable to convince his fellow naturalists that Darwin's ideas were unworthy of serious attention, Agassiz ceased to oppose evolution in the professional forum of intellectual interchange. Instead, he shifted his attack to the popular level, and in this comfortable area where his prestige was still high, he became more dogmatic with the passing of the years. In this way he sharpened the intellectual isolation that the public involvements of the 1850's had created for him and to which he devoted himself so largely during his later career. Among his endeavors were the raising of large sums for the Museum of Comparative Zoology (nearly $600,000 by the time of his death), helping to establish the National Academy of Sciences, aiding the organization of Cornell University, and founding a summer school for the study of natural history. In the years just before his death in 1873 he strove to evaluate the evolutionary concept in a more objective fashion, but his efforts involved no fundamental change in attitude. His popular renown was never dimmed by his determined

[12] In I (1858), 320–323.

anti-Darwinian stand.[13] General appreciation of him is perhaps most clearly expressed by Holmes, who wrote to him a few years after the publication of the *Essay* and the subsequent Darwinian debates:

> It very rarely happens that the same person can take at once the largest and deepest scientific views and come down without apparent effort to the level of popular intelligence. This is what singularly gifts you for our country. . . . You have gained the heart of our purpose . . .[14]

The major professional review of the *Essay* appeared in the *American Journal of Science* and was written by James Dwight Dana[15] a perfect choice for such an assessment because he was a highly competent naturalist who had done much important work in the geographical distribution of marine animals as well as in geology. Though opposed to the hypothesis of development, he believed with Gray that the issue of development versus creation was far from resolved. His overall conclusion was that Agassiz's work "has already borne science to a higher level than it had before attained, and given a force and direction to thought which will insure rapid progress towards perfection." [16] Affirming that Agassiz's views would hardly meet with unqualified acceptance, he praised the "honest purpose in research, thoroughness of investigation, breadth of philosophical ideas, and beauty of actual results" [17] that marked Agassiz's effort. He doubted that Agassiz's taxonomy, representing an attempt to correct certain designations of Cuvier, would prove as useful or valuable as claimed. He objected that some of the larger divisions and a few of the smaller subdivisions were too rigid, especially in the classification of marine animals and the higher vertebrates. This was an understandable criticism in view of Agassiz's effort to have classes, orders, and families in one branch of the animal kingdom correspond in value with those in other branches in a neatly balanced scheme of classification. Such reservations about the inflexible views in the *Essay* somewhat strained Dana's praise, but he agreed with Agassiz's basic belief in the permanence of species and with his insistence that present knowledge made the concept of development a "deluding

[13] For the details of Agassiz's life see Edward Lurie, *Louis Agassiz: A Life in Science* (Chicago, 1960).

[14] Holmes to Agassiz, October 20, 1863, Agassiz Papers, Houghton Library, Harvard University.

[15] In XXV, (2d ser., 1858), 126–128, 202–216, 321–341.

[16] *Ibid.*, p. 341.

[17] *Ibid.*, p. 127.

speculation." Yet he could not admit that Agassiz had said the final word on the problem of species — naturalists, he affirmed, needed to know much more about the causes, mechanisms, and limitations of variation. Nor was he satisfied that Agassiz had proved his contention regarding the independent, plural origin of men and animals. His doubts suggested subsequent acceptance of the evolutionary concept.

Dana's analysis reflected both Agassiz's stature in national science and the restiveness of colleagues who felt too confined by the fetters of special creationism. In his review the Yale naturalist spoke of future volumes by Agassiz in the *Contributions* series as books that would "make still broader the foundation for a true philosophy of nature." But only two more studies — specialized treatments of marine biology — were ever published, a further indication of the totality of Agassiz's public involvements in later years. It thus remained for other men to do what he had urged in the *Essay*:

> As long as men inquire, they will find opportunities to know more upon these topics than those who have gone before them, so inexhaustibly rich is nature in the innermost diversity of her treasures of beauty, order, and intelligence. (p. 141)

A NOTE ON THE TEXT

The present text follows essentially that of the London edition of 1859, in which Agassiz incorporated both additions and corrections to the first American edition of 1857. One such addition, for example, is Section IX of Chapter II on "categories of analogy." Others, perhaps minor, consisted of details and emendations of words and phrases designed to clarify specific arguments. Agassiz also added further documentation in 1859, and where such data seem important to his discussion they have been included here. In both editions the *Essay* was as much a bibliography of the history and current status of zoology in the form of lengthy footnotes as it was a treatise in natural science. This display of sources was important to Agassiz's purpose, because he wanted to show that classical biology rested on firm traditional foundations. His citations, however, were often repetitive and to the modern view not always germane to the subject under discussion. In such instances they have therefore been omitted or shortened. Excessive punctuation has also been eliminated, and, for consistency, correctness, and convenience, many footnotes have been modernized. All editor's notes are enclosed in square brackets. Full references to the history and status of zoological research at the time of the publication of the *Essay* may be found in Agassiz's *Bibliographia Zoologiæ et Geologiæ* (4 vols., London, 1848–1854), edited by H. E. Strickland and Sir William Jardine; and in Wilhelm Engelmann, *Bibliotheca Historico-Naturalis* (Leipzig, 1846).

ESSAY ON CLASSIFICATION

ESSAY ON CLASSIFICATION

CHAPTER I

THE FUNDAMENTAL RELATIONS OF ANIMALS ONE TO ANOTHER AND TO THE WORLD IN WHICH THEY LIVE AS THE BASIS OF THE NATURAL SYSTEM OF ANIMALS

SECTION I

THE LEADING FEATURES OF A NATURAL ZOOLOGICAL SYSTEM ARE ALL FOUNDED IN NATURE

Modern classifications of animals and plants are based upon the peculiarities of their structure; and this is generally considered as the most important, if not the only safe, guide in our attempts to determine the natural relations which exist between animals. This view of the subject seems to me, however, to circumscribe the foundation of a natural system of Zoology and Botany within too narrow limits, to exclude from our consideration some of the most striking characteristics of the two organic kingdoms of nature, and to leave it doubtful how far the arrangement thus obtained is founded in reality, and how far it is merely the expression of our estimate of these structural differences. It has appeared to me appropriate therefore to present here a short exposition of the leading features of the animal kingdom, as an introduction to the study of Natural History in general and of Embryology in particular, as it would afford a desirable opportunity of establishing a standard of comparison between the changes animals undergo during their growth, and the permanent characters of full-grown individuals of other types, and perhaps of showing also what other points beside structure might with advantage be considered in ascertaining the manifold relations of animals to one another and to the world in which they live, upon which the natural system may be founded.

In considering these various topics, I shall of necessity have to discuss many questions bearing upon the very origin of organized beings and to touch upon many points now under discussion among scientific men. I shall, however, avoid controversy as much as possible and only try to render the results of my own studies and meditations in as clear a manner as I possibly can in the short space of an essay like this.

There is no question in Natural History on which more diversified opinions are entertained than on that of Classification; not that naturalists disagree as to the necessity of some sort of arrangement in describing animals or plants, for since nature has become the object of special studies it has been the universal aim of all naturalists to arrange the objects of their investigations in the most natural order possible. Even Buffon,[1] who began the publication of his great Natural History by denying the existence in nature of any thing like a system, closed his work by grouping the birds according to certain general features exhibited in common by many of them. It is true, authors have differed in their estimation of the characters on which their different arrangements are founded; and it is equally true that they have not viewed their arrangements in the same light, some having plainly acknowledged the artificial character of their systems, while others have urged theirs as the true expression of the natural relations which exist between the objects themselves. But, whether systems were presented as artificial or natural, they have to this day been considered generally as the expression of man's understanding of natural objects, and not as a system devised by the Supreme Intelligence and manifested in these objects.[2]

There is only one point in these innumerable systems on which all seem to meet, namely, the existence in nature of distinct species persisting with all their peculiarities, for a time at least; for even

[1] [Georges L. LeClerc de Buffon, 1707–1799.]

[2] The expressions constantly used with reference to genera and species and the higher groups in our systems — as, Mr. A. *has made* such a species *a genus;* Mr. B. *employs* this or that species to *form his genus;* and in which most naturalists indulge when speaking of *their* species, *their* genera, *their* families, *their* systems — exhibit in an unquestionable light the conviction, that such groups are of their own making; which can, however, if the views I shall present below are at all correct, only be true in so far as these groups are *not* true to nature.

the immutability of species has been questioned.[3] Beyond species, however, this confidence in the existence of the divisions, generally admitted in zoological systems, diminishes greatly.

With respect to genera, we find already the number of the naturalists who accept them as natural divisions much smaller; few of them having expressed a belief that genera have as distinct an existence in nature as species. And as to families, orders, classes, or any kind of higher divisions, they seem to be universally considered as convenient devices, framed with the view of facilitating the study of innumerable objects and of grouping them in the most suitable manner. The indifference with which this part of our science is generally treated becomes unjustifiable, considering the progress which Zoology in general has made of late. It is a matter of consequence whether genera are circumscribed in our systematic works within these or those limits; whether families inclose a wider or more contracted range of genera; whether such or such orders are admitted in a class and what are the natural boundaries of classes; as well as how the classes themselves are related to one another, and whether or not all these groups are considered as resting upon the same foundation in nature.

Without venturing here upon an analysis of the various systems of Zoology — the prominent features of which are sufficiently exemplified for my purpose by the systems of Linnæus and Cuvier,[4] which must be familiar to every student of Natural History — it is certainly a seasonable question to ask whether the animal kingdom exhibits only those few subdivisions into orders and genera which the Linnæan system indicates, or whether the classes differ among themselves to the extent which the system of Cuvier would lead us to suppose. Or is, after all, this complicated structure of Classification merely an ingenious human invention which every one may shape as he pleases to suit himself? When we remember that all the works on Natural History admit some system or other of this kind, it is certainly an aim worthy of a true naturalist to ascertain what is the real meaning of all these divisions.

Embryology, moreover, forces the inquiry upon us at every step,

[3] Jean Baptiste Lamarck [1744–1829], *Philosophie zoologique* (2 vols., Paris, 1809; 2d ed., 1830); Baden Powell [1796–1860], *Essays on the Spirit of the Inductive Philosophy* (London, 1855). Compare also Sect. xv below.

[4] [Carolus Linnæus, 1707–1778; Georges Cuvier, 1769–1832;] cf. Chap. III, Sect. III.

as it is impossible to establish precise comparisons between the different stages of growth of young animals of any higher group and the permanent characters of full-grown individuals of other types, without first ascertaining what is the value of the divisions with which we may have to compare embryos. My studies in this department for many years have led me to pay the most careful attention to this subject and to make special investigations for its solution.

Before I proceed any further, however, I would submit one case to the consideration of my reader. Suppose that the innumerable articulated animals, which are counted by tens of thousands, nay, perhaps by hundreds of thousands, had never made their appearance upon the surface of our globe, with one single exception: that, for instance, our Lobster (*Homarus americanus*) were the only representative of that extraordinarily diversified type — how should we introduce that species of animals in our systems? Simply as a genus with one species, by the side of all the other classes with their orders, families, etc., or as a family containing only one genus with one species, or as a class with one order and one genus, or as a class with one family and one genus? And should we acknowledge, by the side of Vertebrata, Mollusks, and Radiata, another type of Articulata, on account of the existence of that one Lobster, or would it be natural to call him by a single name, simply as a species, in contradistinction to all other animals? [5] It was the consideration of this supposed case which led me to the investigations detailed below, which, I hope, may end in the ultimate solution of this apparently inextricable question.

Though what I have now to say about this supposed case cannot be fully appreciated before reading my remarks in the following chapter respecting the character of the different kinds of groups adopted in our systems, it must be obvious that our Lobster, to be what we see these animals are, must have its frame constructed upon that very same plan of structure which it exhibits now; and, if I should succeed in showing that there is a difference between the conception of a plan and the manner of its execution, upon which classes are founded in contradistinction to the types to which they belong,

[5] [These are Agassiz's terms for the four great branches of the animal kingdom, designations adopted from Cuvier's classifications. Common examples are: Vertebrata (reptiles, mammals); Mollusks (snails, squid); Radiata (starfishes, sea lilies); Articulata (worms, insects).]

we might arrive at this distinction by a careful investigation of that single Articulate, as well as by the study of all of them; and we might then recognize its type and ascertain its class characters as fully as if the type embraced several classes, and these classes thousands of species. Secondly, this animal has a form, which no one would fail to recognize; so that, if form can be shown to be characteristic of families, we could thus determine its family. Again: besides the general structure, showing the fundamental relations of all the systems of organs of the body to one another in their natural development, our investigation could be carried into the study of the details of that structure in every part, and thus lead to the recognition of what constitutes everywhere generic characters. Finally: as this animal has definite relations to the surrounding world, as the individuals living at the time bear definite relations to one another, as the parts of their body show definite proportions, and as the surface of the body exhibits a special ornamentation, the specific characters could be traced as fully as if a number of other species were at hand for comparison; and they might be drawn and described with sufficient accuracy to distinguish it at any future time from any other set of species found afterwards, however closely these new species might be allied to it. In this case then we should have to acknowledge a separate branch in the animal kingdom, with a class, a family, and a genus, to introduce one species to its proper place in the system of animals. But the class would have no order, if orders determine the rank, as ascertained by the complication of structure; for, where there is but one representative of a type, there is no room for the question of its superiority or inferiority in comparison to others within the limits of the class, orders being groups subordinate to one another in their class. Yet even in this case, the question of the standing of Articulata, as a type among the other great branches of the animal kingdom, would be open to our investigations; but it would assume another aspect from that which it now presents, as the comparison of Articulata with the other types would then be limited to the Lobster and would lead to a very different result from that to which we may arrive, now that this type includes such a large number of most extensively diversified representatives belonging even to different classes. That such speculations are not idle must be apparent to any one who is aware that, during every period in the history of our globe in past geological ages,[6] the general relations, the numeric proportions, and the relative importance of all the types of the animal kingdom have been ever changing, until their present relations were established. Here then the individuals of one species, as observed while living, simultaneously exhibit characters which, to be expressed satisfactorily and in conformity to what nature tells us, would require the establishment, not only of a distinct species, but also of a distinct genus, a distinct family, a distinct class, a distinct branch. Is not this in itself evidence enough that genera, families, orders, classes, and types have the same foundation in nature as species, and that the individuals living at the time have alone a material existence, they being the bearers, not only of all these different categories of structure upon which the natural system of animals is founded, but also of all the relations which animals sustain to the surrounding world — thus showing that species do not exist in nature in a different way from the higher groups, as is so generally believed?

The divisions of animals according to branch, class, order, family, genus, and species, by which we express the results of our investigations into the relations of the animal kingdom, and which constitute the primary question respecting any system of Zoology seem to me to deserve the consideration of all thoughtful minds. Are these divisions artificial or natural? Are they the devices of the human mind to classify and arrange our knowledge in such a manner as to bring it more readily within our grasp and facilitate further investigations, or have they been instituted by the Divine Intelligence as the categories of his mode of thinking? [7] Have we perhaps thus far been only

[6] A series of classifications of animals and plants, exhibiting each a natural system of the types known to have existed simultaneously during the several successive geological periods, considered singly and without reference to the types of other ages, would show in a strong light the different relations in which the classes, the orders, the families, and even the genera and species, have stood to one another during each epoch. Such classifications would illustrate, in the most impressive manner, the importance of an accurate knowledge of the relative standing of all animals and plants, which can only be inferred from the perusal even of those palæontological works in which fossil remains are illustrated according to their association in different geological formations; for in all these works the remains of past ages are uniformly referred to a system established upon the study of the animals now living, thus lessening the impression of their peculiar combination for the periods under consideration.

[7] It must not be overlooked here that a system may be natural, that is, may agree in every respect with the facts in nature, and yet not be considered by its author as the manifestation of the thoughts of a Creator, but merely as the expression of a fact ex-

the unconscious interpreters of a Divine conception in our attempts to expound nature? And when, in our pride of philosophy, we thought that we were inventing systems of science and classifying creation by the force of our own reason, have we followed only, and reproduced, in our imperfect expressions, the plan whose foundations were laid in the dawn of creation, and the development of which we are laboriously studying — thinking, as we put together and arrange our fragmentary knowledge, that we are introducing order into chaos anew? Is this order the result of the exertions of human skill and ingenuity, or is it inherent in the objects themselves, so that the intelligent student of Natural History is led unconsciously, by the study of the animal kingdom itself, to these conclusions, the great divisions under which he arranges animals being indeed but the headings to the chapters of the great book which he is reading? To me it appears indisputable that this order and arrangement of our studies are based upon the natural, primitive relations of animal life — those systems to which we have given the names of the great leaders of our science who first proposed them being in truth but translations into human language of the thoughts of the Creator. And if this is indeed so, do we not find in this adaptability of the human intellect to the facts of creation,[7a] by which we become instinctively, and, as I have said, unconsciously, the translators of the thoughts of God, the most conclusive proof of our affinity with the Divine Mind? And is not this intellectual and spiritual connection with the Almighty worthy our deepest consideration? If there is any truth in the belief that man is made in the image of God, it is surely not amiss for the philosopher to endeavor, by the study of his own mental operations, to approximate the workings of the Divine Reason, learning from the nature of his own mind better to understand the Infinite Intellect from which it is derived. Such a suggestion may at first sight appear irreverent. But who is the truly humble? He who, penetrating into the secrets of creation, arranges them under a formula which he proudly calls his scientific system? Or he who, in

isting in nature — no matter how — which the human mind may trace and reproduce in a systematic form of its own invention.

[7a] The human mind is in tune with nature, and much that appears as a result of the working of our intelligence is only the natural expression of that preestablished harmony. On the other hand the whole universe may be considered as a school in which man is taught to know himself, and his relations to his fellow beings, as well as to the First Cause of all that exists.

the same pursuit, recognizes his glorious affinity with the Creator, and in deepest gratitude for so sublime a birthright strives to be the faithful interpreter of that Divine Intellect with whom he is permitted, nay, with whom he is intended, according to the laws of his being, to enter into communion?

I confess that this question as to the nature and foundation of our scientific classifications appears to me to have the deepest importance, an importance far greater indeed than is usually attached to it. If it can be proved that man has not invented, but only traced this systematic arrangement in nature, that these relations and proportions which exist throughout the animal and vegetable world have an intellectual, and ideal connection in the mind of the Creator, that this plan of creation, which so commends itself to our highest wisdom, has not grown out of the necessary action of physical laws, but was the free conception of the Almighty Intellect, matured in his thought, before it was manifested in tangible external forms — if, in short, we can prove premeditation prior to the act of creation, we have done once and for ever with the desolate theory which refers us to the laws of matter as accounting for all the wonders of the universe and leaves us with no God but the monotonous, unvarying action of physical forces, binding all things to their inevitable destiny.[8] I think our science has now reached that degree of advancement in which we may venture upon such an investigation.

[8] I allude here only to the doctrines of materialists; but I feel it necessary to add that there are physicists who might be shocked at the idea of being considered as materialists who are yet prone to believe that when they have recognized the laws which regulate the physical world and acknowledged that these laws were established by the Deity, they have explained everything, even when they have considered only the phenomena of the inorganic world, as if the world contained no living beings and as if these living beings exhibited nothing that differed from the inorganic world. Mistaking for a causal relation the intellectual connection observable between serial phenomena, they are unable to perceive any difference between disorder and the free, independent, and self-possessed action of a superior mind, and call mysticism even a passing allusion to the existence of an immaterial principle in animals, which they acknowledge themselves in man. (Powell, *Essays,* pp. 385, 466, 478). I would further remark, that, when speaking of creation in contradistinction with reproduction, I mean only to allude to the difference there is between the regular course of phenomena in nature and the establishment of that order of things, without attempting to explain either; for in whatever manner any state of things which has prevailed for a time upon earth may have been introduced, it is self-evident that its establishment and its maintenance for a determined period are two very different things, however frequently they may be mistaken as identical. It is further of itself plain that the laws which may explain the phenomena of the material world, in contradistinction from the organic, cannot be considered as accounting for

The argument for the existence of an intelligent Creator is generally drawn from the adaptation of means to ends, upon which the Bridgewater treatises,[9] for example, have been based. But this does not appear to me to cover the whole ground, for we can conceive that the natural action of objects upon each other should result in a final fitness of the universe and thus produce an harmonious whole; nor does the argument derived from the connection of organs and functions seem to me more satisfactory, for, beyond certain limits, it is not even true. We find organs without functions, as, for instance, the teeth of the whale, which never cut through the gum, the breast in all males of the class of mammalia; these and similar organs are preserved in obedience to a certain uniformity of fundamental structure, true to the original formula of that division of animal life, even when not essential to its mode of existence. The organ remains, not for the performance of a function, but with reference to a plan,[10] and might almost remind us of what we often see in human structures, when, for instance, in architecture, the same external combi-

---

the existence of living beings, even though these have a material body, unless it be actually shown that the action of these laws implies by their very nature the production of such beings. Life in appropriating the physical world to itself with all its peculiar phenomena exhibits, however, some of its own and of a higher order, which cannot be explained by physical agencies. The circumstance that life is so deeply rooted in the inorganic nature, affords, nevertheless, a strong temptation to explain one by the other; but we shall see presently how fallacious these attempts have been.

[9] [Named for Francis Henry Egerton, 8th Earl of Bridgewater, who left £8,000 for the writing of treatises on the "Power, Wisdom and Goodness of God, as Manifested in the Creation." They included the first eight titles in the following Agassiz note, and the fragment by Babbage.]

Thomas Chalmers, *The Adaptation of External Nature to the Moral and Intellectual Constitution of Man* (2 vols., Glasgow, 1839); John Kidd, *The Adaptation of External Nature to the Physical Condition of Man* (London, 1833); William Whewell, *Astronomy and General Physics Considered with Reference to Natural Theology* (London, 1839); Charles Bell, *The Hand, its Mechanism and Vital Endowments, as Evincing Design* (London, 1833); Peter M. Roget, *Animal and Vegetable Physiology Considered with Reference to Natural Theology* (2 vols., London, 1834); William Buckland, *Geology and Mineralogy Considered with Reference to Natural Theology* (2 vols., London, 1836; 2d ed., 1837); William Kirby, *The History, Habits, and Instincts of Animals . . .* (2 vols., London, 1835); William Prout, *Chemistry, Meteorology, and the Function of Digestion Considered with Reference to Natural Theology* (London, 1834). Compare also, Hercule Strauss-Durkheim, *Théologie de la Nature* (3 vols., Paris, 1852); Hugh Miller, *Footprints of the Creator* (Edinburgh, 1849; 3d ed., with a Memoir of the Author by Louis Agassiz, Boston, 1853); Charles Babbage, *The Ninth Bridgewater Treatise, a Fragment* (2d ed., London, 1838).

[10] The unity of structure of the limbs of club-footed or pinnated animals, in which the fingers are never moved, with those which enjoy the most perfect articulations and freedom of motion exhibits this reference most fully.

nations are retained for the sake of symmetry and harmony of proportion, even when they have no practical object.

I disclaim every intention of introducing in this work any evidence irrelevant to my subject or of supporting any conclusions not immediately flowing from it; but I cannot overlook nor disregard here the close connection there is between the facts ascertained by scientific investigations and the discussions now carried on respecting the origin of organized beings. And though I know those who hold it to be very unscientific to believe that thinking is not something inherent in matter, and that there is an essential difference between inorganic and living and thinking beings, I shall not be prevented by any such pretensions of a false philosophy from expressing my conviction that as long as it cannot be shown that matter or physical forces do actually reason, I shall consider any manifestation of thought as evidence of the existence of a thinking being as the author of such thought, and shall look upon an intelligent and intelligible connection between the facts of nature as direct proof of the existence of a thinking God,[11] as certainly as man exhibits the power of thinking when he recognizes their natural relations.

As I am not writing a didactic work, I will not enter here into a detailed illustration of the facts relating to the various subjects submitted to the consideration of my reader beyond what is absolutely necessary to follow the argument, nor dwell at any length upon the conclusions to which they lead; but simply recall the leading features of the evidence, assuming in the argument a full acquaintance with the whole range of data upon which it is founded, whether derived

[11] I am well aware that even the most eminent investigators consider the task of science at an end, as soon as the most general relations of natural phenomena have been ascertained. To many the inquiry into the primitive cause of their existence seems either beyond the reach of man, or as belonging rather to philosophy than to physics. To these the name of God appears out of place in a scientific work, as if the knowledge of secondary agencies constituted alone a worthy subject for their investigations, and as if nature could teach nothing about its Author. Many, again, are no doubt prevented from expressing their conviction that the world was called into existence and is regulated by an intelligent God, either by the fear of being supposed to share clerical or sectarian prejudices; or because it may be dangerous for them to discuss freely such questions without acknowledging at the same time the obligation of taking the Old Testament as the standard by which the validity of their results is to be measured. Science, however, can only prosper when confining itself within its legitimate sphere; and nothing can be more detrimental to its true dignity than discussions like those which took place at the last [1856] meeting of the German association of naturalists, in Göttingen, and which have since then been carried on in several pamphlets in which bigotry vies with personality and invective.

from the affinities or the anatomical structure of animals, or from their habits and their geographical distribution, from their embryology, or from their succession in past geological ages, and the peculiarities they have exhibited during each,[12] believing as I do that isolated and disconnected facts are of little consequence in the contemplation of the whole plan of creation; and that without a consideration of all the facts furnished by the study of the habits of animals, by their anatomy, their embryology, and the history of the past ages of our globe, we shall never arrive at the knowledge of the natural system of animals.

Let us now consider some of these topics more specially.

SECTION II

SIMULTANEOUS EXISTENCE OF THE MOST DIVERSIFIED TYPES UNDER IDENTICAL CIRCUMSTANCES

It is a fact which seems to be entirely overlooked by those who assume an extensive influence of physical causes upon the very existence of organized beings that the most diversified types of animals and plants are everywhere found under identical circumstances. The smallest sheet of fresh water, every point upon the seashore, every acre of dry land teems with a variety of animals and plants. The narrower the boundaries are which may be assigned as the primitive home of all these beings, the more uniform must be the conditions under which they are assumed to have originated; so uniform, indeed, that in the end the inference would be that the same physical causes could produce the most diversified effects.[13] To concede, on

[12] Many points little investigated thus far by most naturalists, but to which I have of late years paid particular attention, are here presented only in an aphoristic form, as results established by extensive investigations, though unpublished, most of which will be fully illustrated in my following volumes, or in a special work upon the plan of the creation. (See AGASSIZ, "On the Difference between Progressive, Embryonic, and Prophetic Types in the Succession of Organized Beings," *Proceedings,* American Association for the Advancement of Science, II (1850), 432–438.

[13] In order fully to appreciate the difficulty alluded to here, it is only necessary to remember how complicated and at the same time how localized the conditions are under which animals multiply. The egg originates in a special organ, the ovary; it grows there to a certain size, until it requires fecundation, that is, the influence of another living being, or at least of the product of another organ, the spermary, to determine the further development of the germ, which, under the most diversified

the contrary, that these organisms may have appeared in the beginning over a wide area, is to grant, at the same time, that the physical influences under which they existed at first were not so specific as to justify the assumption that these could be the cause of their appearance. In whatever connection, then, the first appearance of organized beings upon earth is viewed, whether it is assumed that they originated within the most limited areas, or over the widest range of their present natural geographical distribution, animals and plants being everywhere diversified to the most extraordinary extent, it is plain that the physical influences under which they subsist cannot logically be considered as the cause of that diversity. In this, as in every other respect, when considering the relations of animals and plants to the conditions under which they live, or to one another, we are inevitably led to look beyond the material facts of the case for an explanation of their existence. Those who have taken another view of this subject, have mistaken the action and reaction which exist everywhere between organized beings, and the physical influences under which they live[14] for a causal or genetic connection, and carried their mistake so far as to assert that these manifold influences could really extend to the production of these beings; not considering how inadequate such a cause would be, and that even

---

conditions, in different species, passes successively through all those changes which lead to the formation of a new perfect being. I then would ask, is it probable that the circumstances under which animals and plants originated for the first time can be much simpler, or even as simple, as the conditions necessary for their reproduction only, after they have once been created? Preliminary, then, to their first appearance, the conditions necessary for their growth must have been provided for, if, as I believe, they were created as eggs, which conditions must have been conformable to those in which the living representatives of the types first produced now reproduce themselves. If it were assumed that they originated in a more advanced stage of life, the difficulties would be still greater, as a moment's consideration cannot fail to show, especially if it is remembered how complicated the structure of some of the animals was which are known to have been among the first inhabitants of our globe. When investigating this subject it is of course necessary to consider the first appearance of animals and plants upon the basis of probabilities only, or even simply upon that of possibilities; as with reference to these first-born, at least, the transmutation theory furnishes no explanation of their existence.

For every species belonging to the first fauna and the first flora which have existed upon earth, special relations, special contrivances must, therefore, have been provided. Now, what would be appropriate for the one would not suit the other, so that, excluding one another in this way, they cannot have originated upon the same point; while within a wider area physical agents are too uniform in their mode of action to have laid the foundation for so many such specific differences as existed between the first inhabitants of our globe.

[14] See below, Sect. XVI.

the action of physical agents upon organized beings presupposes the very existence of those beings.[15] The simple fact that there has been a period in the history of our earth, now well known to geologists,[16] when none of these organized beings as yet existed, and when, nevertheless, the material constitution of our globe, and the physical forces acting upon it, were essentially the same as they are now,[17] shows that these influences are insufficient to call into existence any living being.

Physicists know, indeed, these physical agents more accurately than the naturalists, who ascribe to them the origin of organized beings; let us then ask them, whether the nature of these agents is not specific, whether their mode of action is not specific? They will

[15] A critical examination of this point may dispel much of the confusion which prevails in the discussions relating to the influence of physical causes upon organized beings. That there exist definite relations between animals as well as plants and the mediums in which they live, no one at all familiar with the phenomena of the organic world can doubt; that these mediums and all physical agents at work in nature have a certain influence upon organized beings is equally plain. But before any such action can take place and be felt, organized beings must exist. The problem before us involves, therefore, two questions, the influence of physical agents upon animals and plants already in existence, and the origin of these beings. Granting the influence of these agents upon organized beings to the fullest extent to which it may be traced (see Sect. XVI), there remains still the question of their origin upon which neither argument nor observation has yet thrown any light. But according to some, they originated spontaneously by the immediate agency of physical forces and have become successively more and more diversified by changes produced gradually upon them, by these same forces. Others believe that there exist laws in nature which were established by the Deity in the beginning, to the action of which the origin of organized beings may be ascribed; while according to others, they owe their existence to the immediate intervention of an intelligent Creator. It is the object of the following paragraphs to show that there are neither agents nor laws in nature known to physicists under the influence and by the action of which these beings could have originated; that, on the contrary, the very nature of these beings and their relations to one another and to the world in which they live exhibit thought and can therefore be referred only to the immediate action of a thinking being, even though the manner in which they were called into existence remains for the present a mystery.

[16] Few geologists only may now be inclined to believe that the lowest strata known to contain fossils are not the lowest deposits formed since the existence of organized beings upon earth. But even those who would assume that still lower fossiliferous beds may yet be discovered or may have entirely disappeared by the influence of plutonic agencies (POWELL, *Essays,* p. 424) must acknowledge the fact that everywhere in the lowest rocks known to contain fossils at all there is a variety of them found together. (See Sect. VII.) Moreover, the similarity in the character of the oldest fossils found in different parts of the world goes far, in my opinion, to prove that we actually do know the earliest types of the animal kingdom which have inhabited our globe. This conclusion seems fully sustained by the fact that we find everywhere below this oldest set of fossiliferous beds other stratified rocks in which no trace of organized beings can be found.

[17] See below, Sect. XXI.

all answer that they are. Let us further inquire of them, what evidence there is, in the present state of our knowledge, that at any time these physical agents have produced anything they no longer do produce, and what probability there is that they may ever have produced any organized being? If I am not greatly mistaken, the masters in that department of science will, one and all, answer, none whatever.

But the character of the connections between organized beings and the physical conditions under which they live is such as to display thought;[18] these connections are therefore to be considered as established, determined, and regulated by a thinking being. They must have been fixed for each species at its beginning, while the fact of their permanency through successive generations[19] is further evidence that with their natural relations to the surrounding world were also determined the relations of individuals to one another;[20] their generic as well as their family relations, and every higher grade of affinity;[21] showing, therefore, not only thought, in reference to the physical conditions of existence, but such comprehensive thoughts as would embrace simultaneously every characteristic of each species.

Every fact relating to the geographical distribution of animals and plants might be alluded to in confirmation of this argument, but especially the character of every fauna and every flora upon the surface of the globe. How great the diversity of animals and plants living together in the same region may be can be ascertained by the perusal of special works upon the Zoology and Botany of different countries, or from special treatises upon the geographical distribution of animals and plants.[22] I need, therefore, not enter into further

[18] See below, Sect. XVI.
[19] See below, Sect. XV.
[20] See below, Sect. XVII.
[21] See below, Sect. VI.
[22] L. K. Schmarda, *Die geographische Verbreitung der Thiere* (3 vols., Vienna, 1853); William Swainson, *A Treatise on the Geography and Classification of Animals* (London, 1835); E. A. G. Zimmerman, *Specimen Zoologiæ geographicæ, Quadrupedum domicilia et migrationes sistens* (Leiden, 1777); Alexander von Humboldt, *Essai sur la géographie des plantes* (Paris, 1805) and *Ansichten der Natur* (3d ed., Stuttgart and Tübingen, 1849); Robert Brown, *General Remarks on the Botany of Terra Australis* (London, 1814); Joachim F. Schouw, *Grundzüge einer allgemeinen Pflanzengeographie* (Berlin, 1823); Alphonse de Candolle, *Géographie botanique raisonnée* (2 vols., Paris, 1855).

details upon this subject, especially since it is discussed more fully below.[23]

It might perhaps be urged that animals living together in exceptional conditions and exhibiting structural peculiarities apparently resulting from these conditions, such as the blind fish,[24] the blind crawfish, and the blind insects of the Mammoth Cave in Kentucky, furnish uncontrovertible evidence of the immediate influence of those exceptional conditions upon the organs of vision. If this, however, were the case, how does it happen that that remarkable fish, the *Amblyopsis spelæus,* has only such remote affinities to other fishes? Or were, perhaps, the sum of influences at work to make that fish blind capable also of devising such a combination of structural characters as that fish has in common with all other fishes, with those peculiarities which at the same time distinguish it? Does not, rather, the existence of a rudimentary eye discovered by Dr. J. Wyman in the blind fish show that these animals, like all others, were created with all their peculiarities by the fiat of the Almighty, and this rudiment of eyes left them as a remembrance of the general plan of structure of the great type to which they belong? Or will, perhaps, some one of those naturalists who know so much better than the physicists what physical forces may produce, and that they may produce, and have produced every living being known, explain also to us why subterraneous caves in America produce blind fishes, blind crustacea, and blind insects, while in Europe they produce nearly blind reptiles? If there is no thought in the case, why is it then that this very reptile, the *Proteus anguinus,* forms, with a number of other reptiles living in North America and in Japan, one of the most natural series known in the animal kingdom, every member of which exhibits a distinct grade[25] in the scale? [26]

After we have freed ourselves from the mistaken impression that there may be some genetic connection between physical forces and

[23] See below, Sect. IX.

[24] Jeffries Wyman, "Description of a Blind Fish, from a Cave in Kentucky," *American Journal of Science,* XLV (1843), 94–96, and XVII (2d ser., 1854), 258–261; Agassiz, "Observations on the Blind Fish of the Mammoth Cave," in *ibid.,* XI (2d ser., 1851), 127–128.

[25] See below, Sect. XII.

[26] [Darwin in Chapter V of the *Origin of Species* was very specific in disputing Agassiz on this evidence, employing the example of the eyeless fish as proof of the influence of natural selection through the use and disuse of parts.]

organized beings, there remains a vast field of investigation to ascertain the true relations between both, to their full extent, and within their natural limits.[27] A mere reference to the mode of breathing of different types of animals and to their organs of locomotion, which are more particularly concerned in these relations, will remind every naturalist of how great importance in Classification is the structure of these parts and how much better they might be understood in this point of view, were the different structures of these organs more extensively studied in their direct reference to the world in which animals live. If this had been done, we should no longer call by the same common name of legs and wings organs so different as the locomotive appendages of the insects and those of the birds! We should no longer call lungs the breathing cavity of snails, as well as the air pipes of mammalia, birds, and reptiles! A great reform is indeed needed in this part of our science, and no study can prepare us better for it than the investigation of the mutual dependence of the structure of animals and the conditions in which they live.

SECTION III

REPETITION OF IDENTICAL TYPES UNDER THE MOST DIVERSIFIED CIRCUMSTANCES

As much as the diversity of animals and plants living under identical physical conditions shows the independence of organized beings from the medium in which they dwell, so far as their origin is concerned, so independent do they appear again from the same influences when we consider the fact that identical types occur everywhere upon earth under the most diversified circumstances. If we sum up all these various influences and conditions of existence under the common appellation of cosmic influences, or of physical causes, or of climate in the widest sense of the word, and then look around us for the extreme differences in that respect upon the whole surface of the globe, we find still the most similar, nay identical types (and I allude here under the expression of type to the most diversified acceptations of the word) living normally under their action. There is

[27] See below, Sect. XVI.

no structural difference between the herrings of the Arctic, or those of the Temperate zone, or those of the Tropics, or those of the Antarctic regions; there are not any more between the foxes and wolves of the most distant parts of the globe.[28] Moreover, if there were any, and the specific differences existing between them were insisted upon, could any relation between these differences and the cosmic influences under which they live be pointed out, which would at the same time account for the independence of their structure in general? Or, in other words, how could it be assumed that while these causes would produce specific differences, they would at the same time produce generic identity, family identity, ordinal identity, class identity, typical identity? Identity in everything that is truly important, high, and complicated in the structure of animals, produced by the most diversified influences, while at the same time these extreme physical differences, considered as the cause of the existence of these animals, would produce diversity in secondary relations only! What logic!

Does not all this show, on the contrary, that organized beings exhibit the most astonishing independence of the physical causes under which they live; an independence so great that it can only be understood as the result of a power governing these physical causes as well as the existence of animals and plants, and bringing all into harmonious relations by adaptations which never can be considered as cause and effect?

When naturalists have investigated the influence of physical causes upon living beings, they have constantly overlooked the fact that the features which are thus modified are only of secondary importance in the life of animals and plants, and that neither the plan of their structure nor the various complications of that structure are ever affected by such influences. What, indeed, are the parts of the body which are in any way affected by external influences? Chiefly those which are in immediate contact with the external world, such as the skin, and in the skin chiefly its outer layers, its color, the thickness of the fur, the color of the hair, the feathers, and the scales; then the size of the body and its weight, as far as it is dependent on the quality and quantity of the food; the thickness of the shell of Mollusks,

[28] Innumerable other examples might be quoted which will readily present themselves to professional naturalists; those mentioned above may suffice for my argument.

when they live in waters or upon a soil containing more or less limestone, etc. The rapidity or slowness of the growth is also influenced in a measure by the course of the seasons in different years; so is also the fecundity, the duration of life, etc. But all this has nothing to do with the essential characteristics of animals.

A book has yet to be written upon the independence of organized beings of physical causes, as most of what is generally ascribed to the influence of physical agents upon organized beings ought to be considered as a connection established between them in the general plan of creation.

SECTION IV

UNITY OF PLAN IN OTHERWISE HIGHLY DIVERSIFIED TYPES

Nothing is more striking throughout the animal and vegetable kingdoms than the unity of plan in the structure of the most diversified types. From pole to pole, in every longitude, mammalia, birds, reptiles, and fishes exhibit one and the same plan of structure, involving abstract conceptions of the highest order, far transcending the broadest generalizations of man, for it is only after the most laborious investigations man has arrived at an imperfect understanding of this plan. Other plans, equally wonderful, may be traced in Articulata, in Mollusks, in Radiata,[29] and in the various types of plants. And yet the logical connection, these beautiful harmonies, this infinite diversity in unity are represented by some as the result of forces exhibiting no trace of intelligence, no power of thinking, no faculty of combination, no knowledge of time and space. If there is anything which places man above all other beings in nature, it is precisely the circumstance that he possesses those noble attributes without which, in their most exalted excellence and perfection, not one of these general traits of relationship so characteristic of the great types of the animal and vegetable kingdoms can be understood or even perceived. How, then, could these relations have been devised without similar powers? If all these relations are almost beyond

[29] Agassiz, *Twelve Lectures on Comparative Embryology* (Boston, 1849); "On Animal Morphology," *Proceedings*, AAAS, II (1850), 411–423.

the reach of the mental powers of man, and if man himself is part and parcel of the whole system, how could this system have been called into existence if there does not exist One Supreme Intelligence as the Author of all things?

SECTION V

CORRESPONDENCE IN THE DETAILS OF STRUCTURE IN ANIMALS OTHERWISE ENTIRELY DISCONNECTED

During the first decade of this century naturalists began to study relations among animals which had escaped almost entirely the attention of earlier observers. Though Aristotle knew already that the scales of fishes correspond to the feathers of birds,[30] it is but recently that anatomists have discovered the close correspondence which exists between all the parts of all animals belonging to the same type, however different they may appear at first sight. Not only is the wing of the bird identical in its structure with the arm of man or the fore leg of a quadruped, it agrees quite as closely with the fin of the whale or the pectoral fin of the fish, and all these together correspond in the same manner with their hind extremities. Quite as striking a coincidence is observed between the solid skull-box, the immovable bones of the face and the lower jaw of man and the other mammalia, and the structure of the bony frame of the head of birds, turtles, lizards, snakes, frogs, and fishes. But this correspondence is not limited to the skeleton; every other system of organs exhibits in these animals the same relations, the same identity in plan and structure, whatever be the differences in the form of the parts, in their number, and even in their functions. Such an agreement in the structure of animals is called their homology and is more or less close in proportion as the animals in which it is traced are more or less nearly related.

The same agreement exists between the different systems and their parts in Articulata, in Mollusks, and in Radiata, only that their structure is built up upon respectively different plans, though in these three types the homologies have not yet been traced to the same

[30] Aristoteles, *Historia Animalium,* Lib. I., Chap. 1, Sect. 4 [486^{b}: ". . . for what the feather is in a bird, the scale is in a fish."]

extent as among Vertebrata. There is, therefore, still a wide field open for investigations in this most attractive branch of Zoology. So much, however, is already plain from what has been done in this department of our science, that the identity of structure among animals does not extend to all the four branches of the animal kingdom; that, on the contrary, every great type is constructed upon a distinct plan, so peculiar, indeed, that homologies cannot be extended from one type to the other but are strictly limited within each of them. The more remote resemblance which may be traced between representatives of different types is founded upon analogy and not upon affinity. While, for instance, the head of fishes exhibits the most striking homology with that of reptiles, birds, and mammalia, as a whole, as well as in all its parts, that of Articulata is only analogous to it and to its part. What is commonly called head in Insects is not a head like that of Vertebrata; it has not a distinct cavity for the brain, separated from that which communicates below the neck with the chest and abdomen; its solid envelope does not consist of parts of an internal skeleton, surrounded by flesh, but is formed of external rings, like those of the body, soldered together; it contains but one cavity, which includes the cephalic ganglion, as well as the organs of the mouth and all the muscles of the head. The same may be said of the chest, the legs and wings, the abdomen, and all the parts they contain. The cephalic ganglion is not homologous to the brain, nor are the organs of senses homologous to those of Vertebrata, even though they perform the same functions. The alimentary canal is formed in a very different way in the embryos of the two types, as are also their respiratory organs, and it is as unnatural to identify them, as it would be still to consider gills and lungs as homologous among Vertebrata, now that Embryology has taught us that in different stages of growth these two kinds of respiratory organs exist in all Vertebrata in very different organic connections one from the other.

What is true of the branch of Articulata when compared to that of Vertebrata is equally true of the Mollusks and Radiata when compared with one another or with the two other types, as might easily be shown by a fuller illustration of the correspondence of their structure within these limits. This inequality in the fundamental character of the structure of the four branches of the animal kingdom points to the necessity of a radical reform in the nomenclature of

Comparative Anatomy.[31] Some naturalists, however, have already extended such comparisons respecting the structure of animals beyond the limits pointed out by nature, when they have attempted to show that all structures may be reduced to one norm, and when they have maintained, for instance, that every bone existing in any Vertebrate must have its counterpart in every other species of that type. To assume such a uniformity among animals would amount to denying to the Creator even as much freedom in expressing his thoughts as man enjoys.

If it be true, as pointed out above, that all animals are constructed upon four different plans of structure, in such a manner that all the different kinds of animals are only different expressions of these fundamental formulæ, we may well compare the whole animal kingdom to a work illustrating four great ideas, between which there is no other connecting link than the unity exhibited in the eggs in which their most diversified manifestations are first embodied in an embryonic form, to undergo a series of transformations, and appear in the end in that wonderful variety of independent living beings which inhabit our globe, or have inhabited it from the earliest period of the existence of life upon its surface.

The most surprising feature of the animal kingdom seems, however, to me to rest neither in its diversity, nor in the various degrees of complication of its structure, nor in the close affinity of some of its representatives while others are so different, nor in the manifold relations of all of them to one another and the surrounding world, but in the circumstances that beings endowed with such different and such unequal gifts should nevertheless constitute an harmonious whole, intelligibly connected in all its parts.

[31] Agassiz, "On the Structure and Homologies of Radiated Animals, with Reference to the Systematic Position of the Hydroid Polypi," *Proceedings,* AAAS, II (1850), 389–396.

SECTION VI

VARIOUS DEGREES AND DIFFERENT KINDS OF RELATIONSHIP AMONG ANIMALS

The degrees of relationship existing between different animals are most diversified. They are not only akin as representatives of the same species, bearing as such the closest resemblance to one another; different species may also be related as members of the same genus, the representatives of different genera may belong to the same family, and the same order may contain different families, the same class different orders, and the same type several classes. The existence of different degrees of affinity between animals and plants which have not the remotest genealogical connection, which live in the most distant parts of the world, which have existed in periods long gone by in the history of our earth, is a fact beyond dispute, at least, within certain limits, no longer controverted by well informed observers. Upon what can this be founded? Is it that the retentive capacity of the memory of the physical forces at work upon this globe is such that, after bringing forth a type according to one pattern in the infancy of this earth, that pattern was adhered to under conditions, no matter how diversified, to reproduce at another period something similar, and so on, through all ages, until at the period of the establishment of the present state of things, all the infinitude of new animals and new plants which now crowd its surface should be cast in these four moulds, in such a manner as to exhibit, notwithstanding their complicated relations to the surrounding world, all those more deeply seated general relations which establish among them the different degrees of affinity we may trace so readily in all the representatives of the same type? Does all this really look more like the working of blind forces than like the creation of a reflective mind establishing deliberately all the categories of existence we recognize in nature, and combining them in that wonderful harmony which unites all things into such a perfect system, that even to read it as it is established, or even with all the imperfections of a translation, should be considered as the highest achievement of the maturest genius?

Nothing seems to me to prove more directly and more fully the action of a reflective mind, to indicate more plainly a deliberate consideration of the subject, than the different categories upon which species, genera, families, orders, classes, and branches are founded in nature and manifested in material reality in a succession of individuals, the life of which is limited in its duration to comparatively very short periods. The great wonder in these relations consists in the fugitive character of the bearers of this complicated harmony. For while species persist during long periods, the individuals which represent them are ever changing, one set dying after the other in quick succession. Genera, it is true, may extend over longer periods; families, orders, and classes may even have existed during all periods during which animals have existed at all; but whatever may have been the duration of their existence, at all times these different divisions have stood in the same relation to one another and to their respective branches, and have always been represented upon our globe in the same manner, by a succession of ever renewed and short-lived individuals.

As, however, the second chapter of this work is entirely devoted to the consideration of the different kinds and the different degrees of affinity existing among animals, I will not enter here into any details upon this subject, but simply recall the fact that, in the course of time, investigators have agreed more and more with one another in their estimates of these relations, and built up systems more and more conformable to one another. This result, which is fully exemplified by the history of our science,[32] is in itself sufficient to show that there is a system in nature to which the different systems of authors are successive approximations, more and more closely agreeing with it, in proportion as the human mind has understood nature better. This growing coincidence between our systems and that of nature shows further the identity of the operations of the human and the Divine intellect; especially when it is remembered to what an extraordinary degree many *à priori* conceptions relating to nature

[32] Johann B. Spix, *Geschichte und Beurtheilung aller Systeme in der Zoologie* (Nuremberg, 1811); Cuvier, *Histoire des progrès des sciences naturelles* (4 vols., Paris, 1826), and *Histoire des sciences naturelles . . .* (5 vols., Paris, 1841); Henri de Blainville, *Histoire des sciences de l'organization et de leurs progrès* (3 vols., Paris, 1847); Felix A. Pouchet, *Histoire des sciences naturelles au moyen âge* (Paris, 1853). Compare also Chap. II below.

have in the end proved to agree with the reality, in spite of every objection at first offered by empiric observers.

SECTION VII

SIMULTANEOUS EXISTENCE IN THE EARLIEST GEOLOGICAL PERIODS OF ALL THE GREAT TYPES OF ANIMALS

It was formerly believed by geologists and palæontologists that the lowest animals first made their appearance upon this globe and that they were followed by higher and higher types, until man crowned the series. Every geological museum, representing at all the present state of our knowledge, may now furnish the evidence that this is not the case. On the contrary, representatives of numerous families belonging to all the four great branches of the animal kingdom, are well known to have existed simultaneously in the oldest geological formations.[33] Nevertheless, I well remember when I used to hear the great geologists of the time assert that the Corals were the first inhabitants of our globe, that Mollusks and Articulata followed in order, and that Vertebrates did not appear until long after these. What an extraordinary change the last thirty years have brought about in our knowledge and the doctrines generally adopted respecting the existence of animals and plants in past ages! However much naturalists may still differ in their views regarding the origin, the gradation, and the affinities of animals, they now all know that neither Radiata, nor Mollusks, nor Articulata, have any priority one over the other, as to the time of their first appearance upon earth; and though some still maintain that Vertebrata originated somewhat later, it is universally conceded that they were already in existence toward the end of the first great epoch in the history of our globe. I think it would not be difficult to show upon physiological grounds

[33] Roderick Murchison, *The Silurian System* (London, 1839); Murchison, *Siluria. The History of the Oldest Known Rocks Containing Fossils* (London, 1854); Murchison, Phillippe E. P. de Verneuil, and Alexander von Kaiserling, *The Geology of Russia in Europe, and the Ural Mountains* (2 vols., London, 1845); James Hall, *Palæontology of the State of New York* (2 vols., Albany, 1847–1852); Joachim Barrande, *Système silurien du centre de la Bohème* (2 vols., Prague and Paris, 1852); Adam Sedgwick and Frederick McCoy, *A Synopsis of the Classification of the British Palæozoic Rocks . . .* (London, 1855).

that their presence upon earth dates from as early a period as any of the three other great types of the animal kingdom, since fishes exist wherever Radiata, Mollusks, and Articulata are found together, and the plan of structure of these four great types constitutes a system intimately connected in its very essence. Moreover, for the last twenty years every extensive investigation among the oldest fossiliferous rocks has carried the origin of Vertebrata step by step further back, so that whatever may be the final solution of this vexed question, so much is already established by innumerable facts, that the idea of a gradual succession of Radiata, Mollusks, Articulata, and Vertebrata is forever out of the question. It is proved beyond doubt that Radiata, Mollusca, and Articulata are everywhere found together in the oldest geological formations and that very early Vertebrata are associated with them, to continue together through all geological ages to the present time. This shows that even in those early days of the existence of our globe, when its surface did not yet present those diversified features which it has exhibited in later periods, and which it exhibits in still greater variety now, animals belonging to all the great types now represented upon earth were simultaneously called into existence. It shows further that unless the physical elements then at work could have devised such plans and impressed them upon the material world as the pattern upon which Nature was to build forever afterwards, no such general relations as exist among all animals of all geological periods, as well as among those now living, could ever have existed.

This is not all: every class among Radiata, Mollusks, and Articulata, is known to have been represented in those earliest days, with the exception of the Acalephs[34] and Insects only. It is, therefore, not only the plan of the four great types which must have been adopted then, the manner in which these plans were to be executed, the systems of form under which these structures were to be clothed, even

[34] Acalephs [in modern usage coelenterates, *e.g.*, jellyfish] have been found in the Jurassic Limestone of Solenhofen; their absence in other formations may be owing simply to the extraordinary softness of their body. Insects are known as early as the Carboniferous Formation, and may have existed before. — Since the publication of these remarks I have ascertained that *Millepora* is not a Polyp, but belongs to the Hydroids. It is thus shown that Acalephs have existed in the oldest geological periods, since representatives of the family of *Milleporina* occur in the Silurian rocks.

It remains only to be ascertained now whether all the *Zoantharia tabulata* are as truly Hydroids as the genuine *Milleporina* or not.

the ultimate details of structure which in different genera bear definite relations to those of other genera; the mode of differentiation of species, and the nature of their relations to the surrounding media must likewise have been determined, as the character of the classes is as well defined as that of the four great branches of the animal kingdom, or that of the families, the genera, and the species. Again, the first representatives of each class stand in definite relations to their successors in later periods, and as their order of apparition corresponds to the various degrees of complication in their structure and forms a natural series closely linked together, this natural gradation must have been contemplated from the very beginning.[35] There can be the less doubt upon this point, as man, who comes last, closes in his own cycle a series, the gradation of which points from the very beginning to him as its last term. I think it can be shown by anatomical evidence that man is not only the last and highest among the living beings, for the present period, but that he is the last term of a series beyond which there is no material progress possible upon the plan upon which the whole animal kingdom is constructed, and that the only improvement we may look to upon earth for the future must consist in the development of man's intellectual and moral faculties.[36]

The question has been raised of late how far the oldest fossils known may truly be the remains of the first inhabitants of our globe. No doubt extensive tracts of fossiliferous rocks have been intensely altered by plutonic agencies, and their organic contents so entirely destroyed, and the rocks themselves so deeply metamorphosed, that they resemble now more closely eruptive rocks even than stratified deposits. Such changes have taken place again and again up to comparatively recent periods and upon a very large scale. Yet there are entire continents, North America, for instance, in which the palæozoic rocks have undergone little if any alteration, and where the remains of the earliest representatives of the animal and vegetable kingdoms are as well preserved as in later formations. In such deposits the evidence is satisfactory that a variety of animals belonging to different classes of the great branches of the animal kingdom have existed simultaneously from the beginning; so that the assumption of

[35] [To an evolutionist of course this fact is primary evidence of genetic affiliation.]
[36] Agassiz, *An Introduction to the Study of Natural History* (New York, 1847), p. 57.

a successive introduction of these types upon earth is flatly contradicted by well established and well known facts.[37] Moreover, the remains found in the oldest deposits are everywhere closely allied to one another. In Russia, in Sweden, in Bohemia, and in various other parts of the world, where these oldest formations have been altered upon a more or less extensive scale, as well as in North America, where they have undergone little or no change, they present the same general character, that close correspondence in their structure and in the combination of their families, which shows them to have belonged to contemporaneous faunæ. It would, therefore, seem that even where metamorphic rocks prevail, the traces of the earliest inhabitants of this globe have not been entirely obliterated.

SECTION VIII

THE GRADATION OF STRUCTURE AMONG ANIMALS

There is not only variety among animals and plants; they differ also as to their standing, their rank, their superiority or inferiority when compared to one another. But this rank is difficult to determine; for while in some respects all animals are equally perfect, as they perform completely the part assigned to them in the general economy of nature,[38] in other respects there are such striking differences between them, that their very agreement in certain features points at their superiority or inferiority in regard to others.

This being the case, the question first arises, Do all animals form one unbroken series from the lowest to the highest? Before the animal kingdom had been studied so closely as it has been of late, many able writers really believed that all animals formed but one simple continuous series, the gradation of which Bonnet has been particularly industrious in trying to ascertain.[39] At a later period Lamarck[40]

[37] Agassiz, "The Primitive Diversity and Number of Animals in Geological Times," *American Journal of Science,* XVII (2d ser., 1854), 309–354.

[38] Christian Ehrenberg, *Das Naturreich des Menschen, oder das Reich der willensfreien beseelten Naturkörper, in 29 Classen übersichtlich geordnet* (Berlin, 1835, folio, 1 sheet).

[39] Charles Bonnet, *Considérations sur les corps organisés* (2 vols., Amsterdam, 1762); *Contemplations de la nature* (2 vols., Amsterdam, 1764–1765); *Palingénésie philosophique* (2 vols., Geneva, 1769).

[40] *Philosophie zoologique* (1809).

has endeavored to show further that in the complication of their structure all the classes of the animal kingdom represent only successive degrees, and he is so thoroughly convinced that in his systematic arrangement classes constitute one gradual series, that he actually calls the classes "degrees of organization." De Blainville[41] has in the main followed in the steps of Lamarck, though he does not admit quite so simple a series, for he considers the Mollusks and Articulates as two diverging branches ascending from the Radiata, to converge again and unite in the Vertebrata. But now, since it is known how the great branches of the animal kingdom may be circumscribed,[42] notwithstanding a few doubtful points; since it is still more accurately known how most classes should be characterized and what is their respective standing; since every day brings dissenting views respecting the details of classification nearer together, the supposition that all animals constitute one continuous gradated series can be shown to be contrary to nature. Yet the greatest difficulty in this inquiry is to weigh rightly the respective standing of the

[41] *De l'Organisation des animaux* (Paris, 1822).

[42] Johann Blumenbach, *Handbuch der vegleichenden Anatomie* (Göttingen, 1824; tr., William Lawrence, London, 1827); Cuvier, *Leçons d'anatomie comparée . . .* (5 vols., Paris: G. Duméril et G. Duvernoy, 1800–1805; 2d ed., 10 vols., F. Cuvier, C. Laurillard, et G. Duvernoy, 1836–1844); *Le Règne animal distribué d'apres son organisation* (4 vols., Paris, 1817; 2d ed., 5 vols., A. Latreille, 1829–1830; 3d ed., illust. 8 vols., J. Audouin, *et al.*, 1836–1846; tr., Edward Griffith, *et al.*, 16 vols., London, 1824–1833); Johann F. Meckel, *System der vergleichenden Anatomie* (6 vols., Halle, 1821–1831); Gottfried Treviranus, *Biologie oder der Philosophie der lebenden Natur* (6 vols., Göttingen, 1802–1816), and *Die Erscheinungen und Gesetze des organischen Lebens* (5 vols., Bremen, 1831–1837); Stefano delle Chiaje, *Istituzioni d'Anatomia e Fisiologia comparata* (Naples, 1832); Carl G. Carus, *Lehrbuch der vergleichenden Zootomie* (2d ed., 2 vols., Leipzig, 1834), *Grundsätze der vergleichenden Anatomie* (Dresden, 1828), translated by R. T. Gore as *An Introduction to the Comparative Anatomy of Animals* (2 vols., Bath, 1827), and *Erläuterungstafeln zur vergleichenden Anatomie* (6 vols., Leipzig, 1826–1843); Rudolf Wagner, *Lehrbuch der vergleichenden Anatomie* (2 vols., Leipzig, 1834–1835; tr., A. Tulk, London, 1844); *Lehrbuch der Zootomie* (2 vols., Leipzig, 1843–1845), and *Icones zootomicæ . . .* (Leipzig, 1841); Robert E. Grant, *Outlines of Comparative Anatomy* (London, 1835); T. Rymer Jones, *A General Outline of the Animal Kingdom . . .* (London, 1841; 2d ed., 1854); Robert Todd, *The Cyclopedia of Anatomy and Physiology* (4 vols., London, 1835–1852); Agassiz and Augustus A. Gould, *Principles of Zoology* (Pt. I, Boston, 1848; 2d ed., 1851); Richard Owen, *Lectures on the Comparative Anatomy and Physiology of the Invertebrate Animals* (London, 1843; 2d ed., 1855), and *Lectures on the Comparative Anatomy and Physiology of the Vertebrate Animals, Fishes* (London, 1846); Carl von Siebold and Hermann Stannius, *Lehrbuch der vegleichenden Anatomie* (2 vols., Berlin, 1845; 2d ed., 1855; tr., W. J. Burnett, Boston, 1854); Carl Bergmann and Rudolf Leuckart, *Vergleichende Anatomie und Physiologie* (Stuttgart, 1852); J. V. Carus, *Icones zootomicæ . . .* (Leipzig, 1857); Janus Van der Hoeven, *Handbook of Zoology* (Cambridge, 1856; tr., William Clark).

four great branches of the whole animal kingdom; for, however plain the inferiority of the Radiata may seem when compared with the bulk of the Mollusks or Articulata, or still more evident when contrasted with the Vertebrata, it must not be forgotten that the structure of most Echinoderms is far more complicated than that of any Bryozoon or Ascidian of the type of Mollusks, or that of any Helminth of the type of Articulata, and perhaps even superior to that of the *Amphioxus* among Vertebrata. These facts are so well ascertained, that an absolute superiority or inferiority of one type over the other must be unconditionally denied. As to a relative superiority or inferiority, however determined by the bulk of evidence, though it must be conceded that the Vertebrata rank above the three other types, the question of the relative standing of Mollusks and Articulata seems rather to rest upon a difference in the tendency of their whole organization than upon a real gradation in their structure; concentration being the prominent trait of the structure of Mollusks, while the expression "outward display" would more naturally indicate that of Articulata, and so it might seem as if Mollusks and Articulata were standing on nearly a level with one another, and as much above Radiata as both stand below Vertebrata, but constructed upon plans expressing different tendencies. To appreciate more precisely these most general relations among the great types of the animal kingdom, will require deeper investigations into the character of their plan of structure than have been made thus far.[43] Let, however, the respective standing of these great divisions be what it may; let them differ only in tendency or in plan of structure or in the height to which they rise, admitting their base to be on one level or nearly so; so much is certain, that in each type there are representatives exhibiting a highly complicated structure and others which appear very simple. Now the very fact that such extremes may be traced within the natural boundaries of each type shows that in whatever manner these great types are supposed to follow one another in a single series, the highest representative of the preceding type must join on to the lowest representative of the following, thus bringing necessarily together the most heterogeneous forms.[44] It must be further evident that in proportion as the internal

[43] Compare my paper, "Progressive, Embryonic, and Prophetic Types," *Proceedings*, AAAS, II (1850), 432–438.

[44] Agassiz, "Animal Morphology," *Proceedings*, AAAS, II (1850), 415.

arrangement of each great type will be more perfected, the greater is likely to appear the difference at the two ends of the series which are ultimately to be brought into connection with those of other series, in any attempt to establish a single series for all animals.

I doubt whether there is a naturalist now living who could object to an arrangement in which, to determine the respective standing of Radiata, Polyps would be placed lowest, Acalephs next, and Echinoderms highest; a similar arrangement of Mollusks would bring Acephala lowest, Gasteropoda next, and Cephalopoda highest; Articulata would appear in the following order: Worms, Crustacea, and Insects, and Vertebrata, with the Fishes lowest, next Reptiles and Birds, and Mammalia highest. I have here purposely avoided every allusion to controverted points. Now if Mollusks were to follow Radiata in a simple series, Acephala should join on to the Echinoderms; if Articulata, Worms would be the connecting link. We should then have either Cephalopods or Insects, as the highest term of a series beginning with Radiata, followed by Mollusks or by Articulates. In the first case Cephalopods would be followed by Worms; in the second, Insects by Acephala. Again, the connection with Vertebrata would be made either by Cephalopods, if Articulata were considered as lower than Mollusks, or by Insects, if Mollusks were placed below Articulata. Who does not see, therefore, that in proportion as our knowledge of the true affinities of animals is improving, we accumulate more and more convincing evidence against the idea that the animal kingdom constitutes one simple series?

The next question would then be: Does the animal kingdom constitute several or any number of graduated series? In attempting to ascertain the value of the less comprehensive groups, when compared to one another, the difficulties seem to be gradually less and less. It is already possible to mark out with tolerable precision the relative standing between the classes, though even here we do not yet perceive in all the types the same relations. Among Vertebrata there can be little if any doubt that the Fishes are lower than the Reptiles, these lower than Birds, and that Mammalia stand highest; it seems equally evident that, in the main, Insects and Crustacea are superior to Worms, Cephalopods to Gasteropods, and Acephala and Echinoderms to Acalephs and Polypi. But there are genuine Insects, the superiority of which over many Crustacea would be difficult to prove; there are Worms which in every respect appear superior to certain Crustacea; the structure of the highest Acephala seems more perfect than that of some Gasteropods, and that of the Halcyonoid Polyps more perfect than that of many Hydroids. Classes do, therefore, not seem to be so limited in the range of their characters as to justify in every type a complete serial arrangement among them. But when we come to the orders it can hardly be doubted that the gradation of these natural divisions among themselves in each class constitutes the very essence of this kind of group. As a special paragraph is devoted to the consideration of the character of orders in my next chapter, I need not dwell longer upon this point here.[45] It will be sufficient for me to remark now that the difficulties geologists have met with in their attempts to compare the rank of the different types of animals and plants with the order of their succession in different geological periods has chiefly arisen from the circumstance that they have expected to find a serial gradation, not only among the classes of the same type, where it is only incomplete, but even among the types themselves, between which such a gradation cannot be traced. Had they limited their comparisons to the orders which are really founded upon gradation, the result would have been quite different; but to do this requires more familiarity with Comparative Anatomy, with Embryology and with Zoology proper than can naturally be expected of those, the studies of which are chiefly devoted to the investigation of the structure of our globe.

To appreciate fully the importance of this question of the gradation of animals and to comprehend the whole extent of the difficulties involved in it, a superficial acquaintance with the perplexing question of the order of succession of animals in past geological ages is by no means sufficient. On the other hand, a complete familiarity with the many attempts which have been made to establish a correspondence between the two and with all the crudities which have been published upon this subject might dispel every hope to arrive at any satisfactory result upon this subject, did it not appear now that the inquiry to be conducted upon its true ground must be circumscribed within different limits. The results at which I have already arrived, since I have perceived the mistake under which investigators have been laboring thus far in this respect, satisfy me

[45] See Chap. II, Sect. III.

that the point of view under which I have presented the subject here is the true one; and that in the end the characteristic gradation exhibited by the orders of each class will present the most striking correspondence with the character of the succession of the same groups in past ages, and afford another startling proof of the admirable order and gradation in the degrees of complication of the structure of animals, which have been established from the very beginning and maintained throughout all time.

SECTION IX

RANGE OF THE GEOGRAPHICAL DISTRIBUTION OF ANIMALS

The surface of the earth being partly formed by water and partly by land, and the organization of all living beings standing in close relation to the one or the other of these mediums, it is in the nature of things that no single species, either of animals or plants, should be uniformly distributed over the whole globe. Yet there are some types of the animal as well as of the vegetable kingdom which are equably distributed over the whole surface of the land, and others which are as widely scattered in the sea, while others are limited to some continent or some ocean, to some particular province, to some lake, nay, to some very limited spot of the earth's surface.[46]

As far as the primary divisions of animals are concerned, and the nature of the medium to which they are adapted does not interfere, representatives of the four great branches of the animal kingdom are everywhere found together. Radiata, Mollusks, Articulata, and Vertebrata occur together in every part of the ocean, in the Arctics, as well as under the equator, and near the southern pole as far as man has penetrated; every bay, every inlet, every shoal is haunted by them. So universal is this association, not only at present but in all past geological ages, that I consider it as a sufficient reason to expect

[46] The human race affords an example of the wide distribution of a terrestrial type; the Herring and the Mackerel families have an equally wide distribution in the sea. The Mammalia of New Holland show how some families may be limited to one continent; the family of *Labyrinthici* of the Indian Ocean, how fishes may be circumscribed in the sea, and that of the Goniodonts of South America in the fresh waters. The Chaca of Lake Baikal is found nowhere else. This is equally true of the Blindfish (*Amblyopsis*) of the Mammoth Cave, and of the *Proteus* of the caverns of Carinthia.

that fishes will be found in those few fossiliferous beds of the Silurian System in which thus far they have not yet been found.[47] Upon land we find equally everywhere Vertebrata, Articulata, and Mollusks, but no Radiata, this whole branch being limited to the waters; but as far as terrestrial animals extend, we find representatives of the other three branches associated, as we find them all four in the sea. Classes have already a more limited range of distribution. Among Radiata, the Polypi, Acalephs, and Echinoderms are not only all aquatic, they are all marine, with a single exception,[48] the genus Hydra, which inhabits fresh waters. Among Mollusks, the Acephala are all aquatic, but partly marine and partly fluviatile, the Gasteropoda partly marine, partly fluviatile and partly terrestrial, while all Cephalopoda are marine. Among Articulata, the Worms are partly marine, partly fluviatile, and partly terrestrial, while many are internal parasites, living in the cavities or in the organs of other animals; the Crustacea are partly marine and partly fluviatile, a few are terrestrial; the Insects are mostly terrestrial or rather aërial, yet some are marine, others fluviatile, and a large number of those, which in their perfect state live in the air, are terrestrial or even aquatic during their earlier stages of growth. Among Vertebrata the Fishes are all aquatic, but partly marine and partly fluviatile; the Reptiles are either aquatic or amphibious or terrestrial, and some of the latter are aquatic during the early part of their life; the Birds are all aërial, but some more terrestrial and others more aquatic; finally, the Mammalia, though all aërial, live partly in the sea, partly in fresh water, but mostly upon land. A more special review might show that this localization in connection with the elements in which animals live has a direct reference to peculiarities of structure of such importance, that a close consideration of the habitat of animals within the limits of the classes might in most cases lead to a very natural classification.[49] But this is true only within the limits of the classes, and even here not absolutely, as in some the orders only, or the families only are thus closely related to the elements; there are even natural groups in which this connection is not manifested beyond

[47] See above, Sect. VII.

[48] I need hardly say in this connection that so-called fresh-water Polyps, Alcyonella, Plumatella, etc., are Bryozoa, and not true Polyps.

[49] Agassiz, "The Natural Relations between Animals and the Elements in which They Live," *American Journal of Science,* IX (2d ser., 1850), 369–394.

the limits of the genera, and a few cases in which it is actually confined to the species. Yet in every degree of these connections we find that upon every spot of the globe it extends simultaneously to the representatives of different classes and even of different branches of the animal and vegetable kingdoms; a circumstance which shows that when called into existence in such an association, these various animals and plants were respectively adapted, with all the peculiarities of their kingdom, those of their class, those of their order, those of their genus, and those of their species, to the home assigned to them, and therefore not produced by the nature of the place, or of the element, or any other physical condition.[50] To maintain the contrary would really amount to asserting that wherever a variety of organized beings live together, no matter how great their diversity, the physical agents prevailing there must have in their combined action the power of producing such a diversity of structures as exists in animals, notwithstanding the close connection in which these animals stand to them, or to work out an intimate relation to themselves in beings, the essential characteristics of which have no reference to their nature. In other words, in all these animals and plants there is one side of their organization which has an immediate reference to the elements in which they live, and another which has no such connection, and yet it is precisely this part of the structure of animals and plants which has no direct bearing upon the conditions in which they are placed in nature, which constitutes their essential, their typical character. This proves beyond the possibility of an objection that the elements in which animals and plants live (and under this expression I mean to include all that is commonly called physical agents, physical causes, etc.) cannot in any way be considered as the cause of their existence.

If the naturalists of past centuries have failed to improve their systems of Zoology by introducing considerations derived from the

[50] In the study of the geographical distribution of animals and plants and their relations to the conditions under which they live, too little importance is attached to the circumstance that representations of the most diversified types are everywhere found associated, within limited areas, under identical conditions of existence. These combinations of numerous and most heterogeneous types, under all possible variations of climatic influences, severally circumscribed within the narrowest limits, seems to me to present the most insuperable objection to the supposition that the organized beings, so combined, could in any way have originated spontaneously by the working of any natural law.

habitat of animals, it is chiefly because they have taken this habitat as the foundation of their primary divisions; but reduced to its proper limits, the study of the connection between the structure and the natural home of animals cannot fail to lead to interesting results, among which the growing conviction that these relations are not produced by physical agents, but determined in the plan ordained from the beginning, will not be the least important.

The unequal limitation of groups of a different value upon the surface of the earth produces the most diversified combinations possible, when we consider the mode of association of different families of animals and plants in different parts of the world. These combinations are so regulated that every natural province has a character of its own, as far as its animals and plants are concerned, and such natural associations of organized beings extending over a wider or narrower area are called *Faunæ* when the animals alone are considered, and *Floræ* when the plants alone are regarded. Their natural limits are far from being yet ascertained satisfactorily everywhere. As the works of Schouw and Schmarda may suffice to give an approximate idea of their extent,[51] I would refer to them for further details and allude here only to the unequal extent of these different faunæ and to the necessity of limiting them in different ways, according to the point of view under which they are considered, or rather show that as different groups have a wider or more limited range, in investigating their associations or the faunæ, we must distinguish between zoological realms, zoological provinces, zoological counties, zoological fields, as it were; that is, between zoological areas of unequal value over the widest of which range the most extensive types, while in their smaller and smaller divisions we find more and more limited types, sometimes overlapping one another, sometimes placed side by side, sometimes concentric to one another, but always and everywhere impressing a special character upon some part of a wider area, which is thus made to differ from that of any other part within its natural limits.

These various combinations of smaller or wider areas, equally well defined in different types, has given rise to the conflicting views pre-

[51] I would also refer to a sketch I have published of the Faunæ ("Sketch of the Natural Provinces of the Animal World and Their Relation to the Different Types of Man," in J. C. Nott and George R. Gliddon, *Types of Mankind,* Philadelphia, 1854, accompanied with a map and illustrations, pp. lvii–lxxviii).

vailing among naturalists respecting the natural limits of faunæ; but with the progress of our knowledge these discrepancies cannot fail to disappear. In some respect every island of the Pacific upon which distinct animals are found may be considered as exhibiting a distinct fauna, yet several groups of these islands have a common character which unites them into more comprehensive faunæ, the Sandwich Islands, for instance, compared to the Fejees or to New Zealand. What is true of disconnected islands or of isolated lakes is equally true of connected parts of the mainland and of the ocean.

Since it is well known that many animals are limited to a very narrow range in their geographical distribution, it would be a highly interesting subject of inquiry to ascertain what are the narrowest limits within which animals of different types may be circumscribed, as this would furnish the first basis for a scientific consideration of the conditions under which animals may have been created. The time is passed when the mere indication of the continent whence an animal had been obtained could satisfy our curiosity; and the naturalists who, having an opportunity of ascertaining closely the particular circumstances under which the animals they describe are placed in their natural home, are guilty of a gross disregard of the interest of science when they neglect to relate them. Our knowledge of the geographical distribution of animals would be far more extensive and precise than it is now, but for this neglect. Every new fact relating to the geographical distribution of well-known species is as important to science as the discovery of a new species. Could we only know the range of a single animal as accurately as Alphonse de Candolle has lately determined that of many species of plants, we might begin a new era in Zoology. It is greatly to be regretted that in most works containing the scientific results of explorations of distant countries only new species are described, when the mere enumeration of those already known might have added invaluable information respecting their geographical distribution. The carelessness with which some naturalists distinguish species merely because they are found in distant regions, without even attempting to secure specimens for comparison, is a perpetual source of erroneous conclusions in the study of the geographical distribution of organized beings, not less detrimental to the progress of science than the readiness of others to consider as identical animals and plants which may resemble each other closely, without paying the least regard to their distinct origin, and without even pointing out the differences they may perceive between specimens from different parts of the world. The perfect identity of animals and plants living in very remote parts of the globe has so often been ascertained, and it is also so well known how closely species may be allied and yet differ in all the essential relations which characterize species, that such loose investigations are no longer justifiable.

This close resemblance of animals and plants in distant parts of the world is the most interesting subject of investigation with reference to the question of the unity of origin of animals and to that of the influence of physical agents upon organized beings in general. It appears to me, that, as facts now point distinctly to an independent origin of individuals of the same species in remote regions, or of closely allied species representing one another in distant parts of the world, one of the strongest arguments in favor of the supposition, that physical agents may have had a controlling influence in changing the character of the organic world, is gone for ever.

The narrowest limits within which certain Vertebrata may be circumscribed is exemplified among Mammalia by some large and remarkable species: the Orang-Outangs upon the Sunda Islands, the Chimpanzee and the Gorilla along the western coast of Africa, several distinct species of Rhinoceros about the Cape of Good Hope, and in Java and Sumatra, the Pinchaque and the common Tapir in South America, and the eastern Tapir in Sumatra, the East Indian and the African Elephant, the Bactrian Camel and the Dromedary, the Llamas, and the different kinds of wild Bulls, wild Goats, and wild sheep, etc.; among birds by the African Ostrich, the two American Rheas, the Emeu (*Dromæus*) of New Holland, and the Casuary (*Casuarius galeatus*) of the Indian Archipelago, and still more by the different species of doves confined to particular islands in the Pacific Ocean; among Reptiles, by the *Proteus* of the cave of Adelsberg in Carinthia, by the Gopher (*Testudo Polyphemus* Auct.) of our Southern States; among fishes, by the Blind Fish (*Amblyopsis spelæus*) of the Mammoth Cave. Examples of closely limited Articulata may not be so striking, yet the Blind Crawfish of the Mammoth Cave and the many parasites found only upon or within certain species of animals are very remarkable in this respect. Among Mollusks I would remark the many species of land shells, ascertained by Professor Adams to occur only in Jamaica,[52] among the West India Islands, and the species discovered by the United States Exploring Expedition upon isolated islands of the Pacific and described by Dr. Gould.[53] Even among Radiata many species might be quoted, among Echinoderms as well as among Medusæ and Polypi, which are only known from a few localities; but as long as these animals are not collected with the special view of ascertaining their geographical range, the indications of travelers must be received with great caution, and any generalization respecting the extent of their natural area would be premature as long as the countries they inhabit have not been more extensively explored.[54] It is nevertheless true as established by ample evidence, that within definite limits all the animals occurring in different natural zoological provinces are specifically distinct. What remains to be ascertained more minutely is the precise range of each species, as well as the most natural limits of the different faunæ.[55]

SECTION X

IDENTITY OF STRUCTURE OF WIDELY DISTRIBUTED TYPES

It is not only when considering the diversification of the animal kingdom within limited geographical areas, that we are called upon in our investigations to admire the unity of plan its most diversified types exhibit; the identity of structure of these types is far more surprising when we trace it over a wide range of country and within

[52] Charles B. Adams, *Contributions to Conchology* (12 nos., New York, 1849–1852). A series of pamphlets, full of original information.

[53] Augustus A. Gould, *Mollusca and Shells,* United States Exploring Expedition, *Reports,* XII (Philadelphia, 1852).

[54] With reference to the Echinoderms and Acalephs, I am able to state that the species of the Atlantic shores of North America, found along the northern states, differ entirely from those of the southern states, and these differ again from those of the Gulf of Mexico.

[55] [Agassiz relies heavily on this argument and returns to it time and again, because the concept of the separate and independent creation of species was essential to his denial of common parentage and consequent genetic unity in animals. He therefore had to insist that flora and fauna inhabited limited and distinct zones of creation. An evolutionist would interpret such zones as ecological boundaries that encouraged the variation of species from a common stock.]

entirely disconnected areas. Why the animals and plants of North America should present such a strong resemblance to those of Europe and Northern Asia, while those of Australia are so entirely different from those of Africa and South America under the same latitudes, is certainly a problem of great interest in connection with the study of the influence of physical agents upon the character of animals and plants in different parts of the world. North America certainly does not resemble Europe and Northern Asia, more than parts of Australia resemble certain parts of Africa or of South America; and even if a greater difference should be conceded between the latter than between the former, these disparities are in no way commensurate with the difference or similarity of their organized beings, nor in any way rationally dependent one upon the other. Why should the identity of species prevailing in the Arctics not extend to the temperate zone, when many species of this zone, though different, are as difficult to distinguish as it is difficult to prove the identity of certain arctic species in the different continents converging to the north, and when besides, those of the two zones mingle to a great extent at their boundaries? Why are the antarctic species not identical with those of the arctic regions? And why should a further increase of the average temperature introduce such completely new types, when even in the Arctics, there are in different continents such strikingly peculiar types (*Rhytina* for instance) combined with those that are identical over the whole arctic area? [56]

It may at first sight seem very natural that the arctic species should extend over the three northern continents converging toward the north pole, as there can be no insuperable barrier to the widest dissemination over this whole area for animals living in a glacial ocean or upon parts of three continents which are almost bound together by ice. Yet the more we trace this identity in detail, the more surprising does it appear, as we find in the Arctics as well as every-

[56] I beg not to be misunderstood. I do not impute to all naturalists the idea of ascribing all the differences or all the similarities of the organic world to climatic influences; I wish only to remind them that even the truest picture of the correlations of climate and geographical distribution does not yet touch the question of origin, which is the point under consideration. Too little attention has thus far been paid to the facts bearing upon the peculiarities of structure of animals in connection with the range of their distribution. Such investigations are only beginning to be made as native investigators are studying comparatively the anatomy of animals of different continents.

where else representatives of different types living together. The arctic Mammalia belonging chiefly to the families of Whales, Seals, Bears, Weasels, Foxes, Ruminants and Rodents, have, as Mammalia, the same general structure as the Mammalia of any other part of the globe, and so have the arctic Birds, the arctic Fishes, the arctic Articulata, the arctic Mollusks, the arctic Radiata when compared to the representatives of the same types all over our globe. This identity extends to every degree of affinity among these animals and the plants which accompany them; their orders, their families, and their genera, as far as they have representatives elsewhere, bear everywhere the same identical ordinal, family, or generic characters. The arctic foxes have the same dental formula, the same toes and claws, in fact, every generic peculiarity which characterizes foxes, whether they live in the Arctics, or in the temperate or tropical zone, in America, in Europe, in Africa, or in Asia. This is equally true of the seals or the whales; the same details of structure which characterize their genera in the Arctics reappear in the Antarctics and the intervening space, as far as their natural distribution goes. This is equally true of the birds, the fishes, etc., etc. And let it not be supposed that it is only a general resemblance. By no means. The structural identity extends to the most minute details in the most intimate structure of the teeth, of the hair, of the scales, in the furrows of the brain, in the ramification of the vessels, in the folds of the internal surface of the intestine, in the complication of the glands, etc., etc., to peculiarities, indeed, which nobody but a professional naturalist, conversant with microscopic anatomy, would ever believe could present such precise and permanent characters. So complete, indeed, is this identity, that were any of these beings submitted to the investigation of a skilful anatomist, after having been mutilated to such an extent that none of its specific characters could be recognized, yet not only its class, or its order, or its family, but even its genus, could be identified as precisely as if it were perfectly well preserved in all its parts. Were the genera few which have a wide range upon the earth and in the ocean, this might be considered as an extraordinary case; but there is no class of animals and plants which does not contain many genera, more or less cosmopolite in their geographical distribution. The number of animals which have a wide distribution is even so great that, as far at least as genera are concerned, it may fairly be said that the majority of them have an extensive geographical range. This amounts to the most complete evidence that, as far as any of these genera extends in its geographical distribution, animals the structure of which is identical within this range of distribution are entirely beyond the influence of physical agents, unless these agents have the power, notwithstanding their extreme diversity, within these very same geographical limits, to produce absolutely identical structures of the most diversified types.[57]

It must be remembered here that there are genera of Vertebrata, of Articulata, of Mollusks, and of Radiata which occupy the same identical and wide geographical distribution, and that while the structure of their respective representatives is identical over the whole area, as Vertebrata, as Articulata, as Mollusks, as Radiata, they are at the same time built upon the most different plans. I hold this fact to be in itself a complete demonstration of the entire independence of the structure of animals of physical agents, and I may add that the vegetable kingdom presents a series of facts identical with these. This proves that all the higher relations among animals and plants are determined by other causes than mere physical influences.

While all the representatives of the same genus are identical in structure,[58] the different species of one genus differ only in their size, in the proportions of their parts, in their ornamentation, in their relations to the surrounding elements, etc. The geographical range of these species varies so greatly that it cannot afford in itself a criterion for the distinction of species. It appears further that while

[57] An example may serve to bring this argument nearer to those not familiar with Natural History. From the Arctic Ocean to Cape Horn, America embraces such a variety of physical features that we may well suppose all the natural causes to which the origin of organized beings could be ascribed to be or to have been active within this range. Now there is a peculiar kind of fox in Arctic America; others occur in the temperate zone of that continent, and others again in more southern latitudes. With them the most diversified animals of every class are associated, among which there are many types, the geographical range of which is circumscribed with the narrowest limits; although a large number of them have representatives in other parts of the world. It is plain, therefore, that physical agents cannot be the cause of the existence of any of them, unless these agents act with discrimination, producing mammalia of the same genus over the whole continent, and by the side of them other animals belonging to the most diversified types and agreeing with the extra-American representatives of these types in every essential feature. This is tantamount to assuming that such an action is the work of a rational being.

[58] See Chap. II, Sect. v.

some species which are scattered over very extensive areas occupy disconnected parts of that area, other species closely allied to one another and which are generally designated under the name of representative species occupy respectively such disconnected sections of these areas. The question then arises, how these natural boundaries assigned to every species are established. It is now generally believed that each species had in the beginning some starting point from which it has spread over the whole range of the area it now occupies, and that this starting point is still indicated by the prevalence or concentration of such species in some particular part of its natural area, which, on that account, is called its centre of distribution or centre of creation, while at its external limits the representatives of such species thin out, as it were, occurring more sparsely and sometimes in a reduced condition.

It was a great progress in our science when the more extensive and precise knowledge of the geographical distribution of organized beings forced upon its cultivators the conviction that neither animals nor plants could have originated upon one and the same spot upon the surface of the earth and hence have spread more and more widely until the whole globe became inhabited. It was really an immense progress which freed science from the fetters of an old prejudice. For now we have the facts of the case before us, it is really difficult to conceive how, by assuming such a gradual dissemination from one spot, the diversity which exists in every part of the globe could ever have seemed to be explained. But even to grant distinct centres of distribution for each species within their natural boundaries is only to meet the facts half way, as there are innumerable relations between the animals and plants which we find associated everywhere, which must be considered as primitive, and cannot be the result of successive adaptation. And if this be so, it would follow that all animals and plants have occupied from the beginning those natural boundaries within which they stand to one another in such harmonious relations.[59] Pines have originated in forests, heaths in heathers, grasses in prairies, bees in hives, herrings in schools, buffaloes in herds, men in nations.[60] I see a striking proof that this must

[59] Agassiz, "Geographical Distribution of Animals," *Christian Examiner,* XLVIII (1850), 181–204.

[60] Agassiz, "The Diversity of Origin of the Human Races," *ibid.,* XLIX (1850), 110–145.

have been the case in the circumstance, that representative species, which, as distinct species, must have had from the beginning a different and distinct geographical range, frequently occupy sections of areas which are simultaneously inhabited by the representatives of other species, which are perfectly identical over the whole area.[61] By way of an example, I would mention the European and the American Widgeon (*Anas Penelope* and *A. americana*), or the American and the European Red-headed Ducks (*A. ferina* and *A. erythrocephala*), which inhabit respectively the northern parts of the Old and New World in summer and migrate further south in these same continents during winter, while the Mallard (*A. Boschas*) and the Scaup Duck (*A. marila*) are as common in North America as in Europe. What do these facts tell? That all these birds originated together somewhere where they no longer occur, to establish themselves in the end within the limits they now occupy? — or that they originated either in Europe or America, where, it is true, they do not live all together, but at least a part of them? — or that they really originated within the natural boundaries they occupy? I suppose with sensible readers I need only argue the conclusions flowing from the last supposition. If so, the American Widgeon and the American Red-headed Duck originated in America, and the European Widgeon and the European Red-headed Duck in Europe. But what of the Mallard and the Scaup, which are equally common upon the two continents? Did they first appear in Europe, or in America, or simultaneously upon the two continents? Without entering into further details, as I have only desired to lay clearly a distinct case before my readers from which the character of the argument, which applies to the whole animal kingdom, may be fully understood — I say that the facts lead, step by step, to the inference, that such birds as the Mallard and the Scaup originated simultaneously and separately in Europe and in America and that all animals originated in vast numbers, indeed, in the average number characteristic of their species, over the whole of their geographical area, whether its surface be continuous or disconnected by sea, lakes, or rivers, or by differences of level above the sea, etc. The details of the geographical distribu-

[61] [See Agassiz, "Prefatory Remarks," in Nott and Gliddon, *Indigenous Races of the Earth* (Philadelphia, 1857), pp. xiii–xv, and *A Journey in Brazil* (Boston, 1868), pp. 529–532, for elaborations upon the concept of the plural origins of animal forms with particular reference to man.]

tion of animals exhibit, indeed, too much discrimination to admit for a moment that it could be the result of accident; that is, the result of the accidental migrations of the animals or of the accidental dispersion of the seeds of plants. The greater the uniformity of structure of these widely distributed organized beings, the less probable does their accidental distribution appear. I confess that nothing has ever surprised me so much as to see the perfect identity of the most delicate microscopic structures of animals and plants, from the remotest parts of the world. It was this striking identity of structure in the same types, this total independence of the essential characteristics of animals and plants, of their distribution under the most extreme climatic differences known upon our globe, which led me to distrust the belief, then almost universal, that organized beings are influenced by physical causes to a degree which may essentially modify their character.

SECTION XI

COMMUNITY OF STRUCTURE AMONG ANIMALS LIVING IN THE SAME REGIONS

The most interesting result of the earliest investigations of the fauna of Australia was the discovery of a type of animals, the Marsupialia, prevailing upon this continental island, which are unknown in almost every other part of the world. Every student of Natural History knows now that there are no *Quadrumana* in New Holland, neither Monkeys, nor Makis; no *Insectivora,* neither Shrews, nor Moles, nor Hedgehogs; no true *Carnivora,*[62] neither Bears, nor Weasels, nor Foxes, nor Viverras, nor Hyenas, nor Wild Cats; no *Edentata,* neither Sloths, nor Tatous, nor Ant-eaters, nor Pangolins; no *Pachyderms,* neither Elephants, nor Hippopotamuses, nor Hogs, nor Rhinoceroses, nor Tapirs, nor Wild Horses; no *Ruminantia,* neither Camels, nor Llamas, nor Deers, nor Goats, nor Sheep, nor Bulls, etc., and yet the Mammalia of Australia are almost as diversified as those of any other continent. In the words of Waterhouse,[63] who has

[62] Doubts are entertained respecting the origin of the Dingo, the only beast of prey of New Holland.

[63] George R. Waterhouse, *A Natural History of the Mammalia* (2 vols., London, 1848), I, 4.

studied them with particular care, "the Marsupialia present a remarkable diversity of structure, containing herbivorous, carnivorous, and insectivorous species; indeed, we find amongst the marsupial animals analogous representations of most of the other orders of Mammalia. The *Quadrumana* are represented by the Phelangers, the *Carnivora* by the Dasyuri, the *Insectivora* by the small Phascogales, the *Ruminantia* by the Kangaroos, and the *Edentata* by the Monotremes. The Cheiroptera are not represented by any known marsupial animals, and the Rodents are represented by a single species only; the hiatus is filled up, however, in both cases, by placental species, for Bats and Rodents are tolerably numerous in Australia, and, if we except the Dog, which it is probable has been introduced by man, these are the only placental Mammalia found in that continent." Nevertheless, all these animals have in common some most striking anatomical characters which distinguish them from all other Mammalia and stamp them as one of the most natural groups of that class; their mode of reproduction and the connection of the young with the mother are different; so also is the structure of their brain, etc.[64]

Now the suggestion that such peculiarities could be produced by physical agents is forever set aside by the fact that neither the birds nor the reptiles, nor, indeed, any other animals of New Holland depart in such a manner from the ordinary character of their representatives in other parts of the world; unless it could be shown that such agents have the power of discrimination and may produce, under the same conditions, beings which agree and others which do not agree with those of different continents; not to speak again of the simultaneous occurence in that same continent of other heterogenous types of Mammalia, Bats and Rodents, which occur there as well as everywhere else in other continents.[65] Nor is New Holland the only part of the world which nourishes animals highly diversified among themselves and yet presenting common characters strikingly different from those of the other members of their type, circumscribed within definite geographical areas. Almost every part of the globe exhibits some such group either of animals or of plants, and

[64] See Owen, "Marsupialia," in R. B. Todd, *Cyclopedia of Anatomy and Physiology* (4 vols., 1835–1852), and several elaborate papers by himself and others, quoted there.

[65] [In Chapter IV of the *Origin of Species* Darwin cited Waterhouse's findings as evidence of divergence of character fostered by natural selection.]

every class of organized beings contains some native natural group, more or less extensive, more or less prominent, which is circumscribed within peculiar geographical limits.

Among Mammalia we might quote further the Quadrumana, the representatives of which, though greatly diversified in the Old as well as in the New World, differ and agree respectively in many important points of their structure; also the Edentata of South America. Among birds, the Humming Birds, which constitute a very natural, beautiful, and numerous family, all of which are nevertheless confined to America only, as the Pheasants are to the Old World.[66] Among Reptiles, the Crocodiles of the Old World compared to those of America. Among Fishes, the family of *Labyrinthici,* which is confined to the Indian and Pacific Oceans, that of Goniodonts, which is limited to the fresh waters of South America, as that of Cestracionts to the Pacific. The comparative anatomy of Insects is not sufficiently far advanced to furnish striking examples of this kind. Among Insects, however, remarkable for their form, which are limited to particular regions, may be quoted the genus *Mormolyce* of Java, the *Pneumora* of the Cape of Good Hope, the *Belostoma* of North America, the *Fulgora* of China, etc. The geographical distribution of Crustacea has been treated in such a masterly manner by Dana, in his great work upon the Crustacea of the United States Exploring Expedition, Vol. XIII., p. 1451,[67] that I can only refer to it for numerous examples of localized types of this class, and also as a model how to deal with such subjects. Among Worms, the genus *Peripates* of Guiana deserves to be mentioned. Among Cephalopods, the *Nautilus* in Amboyna. Among Gasteropods, the genus *Io* in the western waters of the United States. Among Acephala, the *Trigonia* in New Holland, certain Naiades in the United States, the *Aetheria* in the Nile. Among Echinoderms, the *Pentacrinus* in the West Indies, the *Culcita* in Zanzibar, the *Amblypneustes* in the Pacific, the *Temnopleurus* in the Indian Ocean, the Dendraster on the western coast of North America. Among Acalephs, the *Berenice* of New Holland. Among Polypi, the true *Fungidæ* in the Indian and Pacific Oceans, the *Renilla* in the Atlantic, etc.

[66] What are called Pheasants in America do not even belong to the same family as the eastern Pheasants. The American so-called Pheasants are generally Grouses.

[67] [James Dwight Dana, *Crustacea,* United States Exploring Expedition, *Reports,* XIII, Philadelphia, 1852.]

Many more examples might be quoted, were our knowledge of the geographical distribution of the lower animals more precise. But these will suffice to show that whether high or low, aquatic or terrestrial, there are types of animals remarkable for their peculiar structure which are circumscribed within definite limits, and this localization of special structures is a striking confirmation of the view expressed already in another connection, that the organization of animals, whatever it is, may be adapted to various and identical conditions of existence, and can in no way be considered as originating from these conditions.

SECTION XII

SERIAL CONNECTION IN THE STRUCTURE OF ANIMALS WIDELY SCATTERED UPON THE SURFACE OF OUR GLOBE

Ever since I have become acquainted with the reptiles inhabiting different parts of the world, I have been struck with a remarkable fact, not yet noticed by naturalists, as far as I know, and of which no other class exhibits such striking examples. This fact is that among Saurians, as well as among Batrachians, there are families, the representatives of which, though scattered all over the globe, form the most natural connected series, in which every link represents one particular degree of development. The Scincoids,[68] among Saurians, are one of these families. It contains about one hundred species, referred by Duméril and Bibron to thirty-one genera, which, in the development of their organs of locomotion, exhibit most remarkable combinations, illustrated in a diagram, on the following page[s].

Fully to appreciate the meaning of this diagram, it ought to be remembered that the animals belonging to this family are considered here in two different points of view. In the first place, their zoological relations to one another are expressed by the various combinations of the structure of their legs; some having four legs, and these are the most numerous, others only two legs, which are always the hind legs, and others still no legs at all. Again these legs may have

[68] For the characters of the family see André M. C. Duméril et G. Bibron, *Erpétologie générale* . . . (9 vols., Paris, 1834–1854), V, 511.

only one toe, or two, three, four, or five toes, and the number of toes may vary between the fore and hind legs. The classification adopted here is based upon these characters. In the second place, the geographical distribution is noticed. But it is at once apparent that the home of these animals stands in no relation whatsoever to their zoological arrangement. On the contrary, the most remote genera may occur in the same country, while the most closely related may live far apart.

GENERA WITH FOUR LEGS

With *five* toes to the fore feet, as well as to the hind feet:	*Tropidophorus,* 1 species, Cochin-China. *Scincus,* 1 sp., Syria, North and West Africa. *Sphenops,* 1 sp., Egypt. *Diploglossus,* 6 sp., West Indies and Brazils. *Amphiglossus,* 1 sp., Madagascar. *Gongylus,* with 7 sub-genera: *Gongylus,* 2 sp., Southern Europe, Egypt, Teneriffe, Isle de France. *Eumeces,* 11 sp., East and West Indies, South America, Vanikoro, New Ireland, New Guinea, Pacific Islands. *Euprepes,* 13 sp., West coast of Africa, Cape of Good Hope, Egypt, Abyssinia, Seychelles, Madagascar, New Guinea, East Indies, Sunda Islands, Manila. *Plestiodon,* 5 sp., Egypt, Algiers, China, Japan, United States. *Lygosoma,* 19 sp., New Holland, New Zealand, Java, New Guinea, Timor, East Indies, Pacific Islands, United States. *Leiolopisma,* 1 sp., Mauritius and Manila. *Tropidolopisma,* 1 sp., New Holland. *Cyclodus,* 3 sp., New Holland and Java. *Trachysaurus,* 1 sp., New Holland. *Ablepharus,* 4 sp., Southeastern Europe, New Holland, Pacific Islands.

With *five* toes to the fore feet and *four* toes to the hind feet: *Campsodactylus,* 1 sp., Bengal.

With *four* toes to the fore feet and *five* toes to the hind feet:	*Heteropus,* 3 sp., Africa, New Holland, Isle de France. *Gymnophthalmus,* 1 sp., W. Indies and Brazil.
With *four* toes to the fore feet and *four* toes to the hind feet:	*Tetradactylus,* 1 sp., New Holland. The genus Chalcides of the allied family Chalcidioids, exhibits another example of this combination.

With *four* toes to the fore feet and *three* toes to the hind feet: No examples known of this combination.

With *three* toes to the fore feet and *four* toes to the hind feet: Not known.

With *three* toes to the fore feet and *three* toes to the hind feet:	*Hemiergis,* 1 sp., New Holland. *Seps,* 1 sp., S. Europe and N. Africa. *Nessia,* 1 sp., Origin unknown.

With *three* toes to the fore feet and *two* toes to the hind feet: Not known.

With *two* toes to the fore feet and *three* toes to the hind feet: { *Heteromeles,* 1 sp., Algiers. *Lerista,* 1 sp., New Holland.

With *two* toes to the fore feet and *two* toes to the hind feet: *Chelomeles,* 1 sp., New Holland.

With *two* toes to the fore feet and *one* toe to the hind feet: *Brachymeles,* 1 sp., Philippine Islands.

With *one* toe to the fore feet and *two* toes to the hind feet: *Brachystopus,* 1 sp., South Africa.

With *one* toe to the fore feet and *one* toe to the hind feet: *Evesia,* 1 sp., Origin unknown.

GENERA WITH ONLY TWO LEGS

No representatives are known *with fore legs only;* but this structural combination occurs in the allied family of the Chalcidioids. The representatives *with hind legs only,* present the following combinations: —

With *two* toes: *Scelotes,* 1 sp., Cape Good Hope.
With *one* toe: *Propeditus,* 1 sp., Cape Good Hope and New Holland.
Ophiodes, 1 sp., South America.
Hysteropus, 1 sp., New Holland.
Lialis, 1 sp., New Holland.
Dibamus, 1 sp., New Guinea.

GENERA WITHOUT ANY LEGS

Anguis, 1 sp., Europe, Western Asia, Northern Africa.
Ophiomorus, 1 sp., Morea, Southern Russia, and Algiers.
Acontias, 1 sp., Southern Africa, Cape Good Hope.
Typhlina, 1 sp., Southern Africa, Cape Good Hope.

Who can look at this diagram, and not recognize in its arrangement the combinations of thought? This is so obvious, that while considering it one might almost overlook the fact that while it was drawn up to classify animals preserved in the Museum of the Jardin des Plantes in Paris, it is in reality inscribed in Nature by these animals themselves and is only read off when they are brought together and compared side by side. But it contains an important element for our discussion: the series is not built up of equivalent representatives in its different terms, some combinations being richly endowed, others numbering a few, or even a single genus, and still others being altogether disregarded; such freedom indicates selection, and not the working of the law of necessity.[69]

And if, from a contemplation of this remarkable series we turn our attention to the indications relating to the geographical distri-

[69] [The foregoing is a clear example of the way in which Agassiz interpreted primary evidence for evolution as proof for special creationism. This very example of leg and toe reduction in lizards is still employed in textbooks in zoology to demonstrate the workings of evolution. See Ernst Mayr, "Agassiz, Darwin, and Evolution," *Harvard Library Bulletin,* XIII (1959), 184.]

bution of these so closely linked genera, inscribed after their names, we perceive at once that they are scattered all over the globe, but not so that there could be any connection between the combinations of their structural characters and their homes. The types without legs are found in Europe, in Western Asia, in Northern Africa, and at the Cape of Good Hope; the types with hind legs only, and with one single toe, at the Cape of Good Hope, in South America, New Holland, and New Guinea; those with two toes at the Cape of Good Hope only. Among the types with four legs the origin of those with but one toe to each foot is unknown, those with one toe in the fore foot and two in the hind foot are from South Africa, those with two toes in the fore foot and one in the hind foot occur in the Philippine Islands, those with two toes to all four feet in New Holland, those with three toes to the hind feet and two to the fore feet in Algiers and New Holland; none are known with three toes to the fore feet and two to the hind feet. Those with three toes to the fore feet inhabit Europe, Northern Africa, and New Holland. There are none with three and four toes, either in the fore feet or in the hind feet. Those with four toes to the fore feet live in New Holland; those with five toes to the fore feet and four to the hind feet, in Bengal, and with four toes in the fore feet and five in the hind feet, in Africa, the West Indies, the Brazils, and New Holland. Those with five toes to all four feet have the widest distribution, and yet they are so scattered that no single zoological province presents any thing like a complete series. On the contrary, the mixture of some of the representatives with perfect feet with others which have them rudimentary, in almost every fauna, excludes still more decidely the idea of an influence of physical agents upon this development.

Another similar series, not less striking, may be traced among the Batrachians, for the characters of which I may refer to the works of Holbrook, Tschudi, and Baird,[70] even though they have not presented them in this connection, as the characteristics of the genera will of themselves suggest their order, and further details upon this subject would be superfluous for my purpose, the more so, as I have already discussed the gradation of these animals elsewhere.[71]

[70] John E. Holbrook, *North American Herpetology* (5 vols., Philadelphia, 1842–1843); Johann J. von Tschudi, *Classification der Batrachier* (Neuchâtel, 1838); Spencer F. Baird, "Revision of the North American Tailed Batrachia," *Journal,* Academy of Natural Sciences of Philadelphia, I (2d ser., 1849), 281–294.

[71] Agassiz, *Twelve Lectures on Embryology,* p. 8.

Similar series, though less conspicuous and more limited, may be traced in every class of the animal kingdom, not only among the living types, but also among the representatives of past geological ages, which adds to the interest of such series in showing that the combinations include not only the element of space, indicating omnipresence, but also that of time, which involves prescience. The series of Crinoids, that of Brachiopods through all geological ages, that of the Nautiloids, that of Ammonitoids from the Trias to the Cretaceous formation inclusive, that of Trilobites from the lowest beds up to the Carboniferous period, that of Ganoids through all formations; then again among living animals in the class of Mammalia, the series of Monkeys in the Old World especially, that of Carnivora from the Seals, through the Plantigrades, to the Digitigrades; in the class of Birds, that of the Wading Birds, and that of the Gallinaceous Birds; in the class of Fishes, that of Pleuronectidæ and Gadoids, that of Skates and Sharks; in the class of Insects, that of Lepidoptera from the Tineina to the Papilionina; in the class of Crustacea, that of the Decapods in particular; in the class of Worms, that of the Nudibranchiata or that of the Dorsibranchiata especially; in the class of Cephalopoda, that of the Sepioids; in the class of Gasteropoda, that of the Nubibranchiata in particular; in the class of Acephala, that of the Ascidians and that of the Oysters in the widest sense; in the class of Echinoderms, those of Holothuriæ and Asterioids; in the class of Acalephs, that of the Hydroids; in the class of Polyps, that of the Halcyonoids, of the Atræoids, etc., etc., deserve particular attention, and may be studied with great advantage in reference to the points under consideration. For everywhere do we observe in them, with reference to space and to time, the thoughtful combinations of an active mind. But it ought not to be overlooked that, while some types represent strikingly connected series, there are others in which nothing of the kind seems to exist and the diversity of which involves other considerations.

SECTION XIII

RELATION BETWEEN THE SIZE OF ANIMALS AND THEIR STRUCTURE

The relation between the size and structure of animals has been very little investigated, though even the most superficial survey of the animal kingdom may satisfy any one that there is a decided relation between size and structure among them. Not that I mean to assert that size and structure form parallel series, or that all animals of one branch or even those of the same class or the same order agree very closely with one another in reference to size. This element of their organization is not defined within those limits, though the Vertebrata, as a whole, are larger than either Articulata, Mollusks, or Radiata; though Mammalia are larger than Birds, Crustacea larger than Insects; though Cetacea are larger than Herbivora, these larger than Carnivora, etc. The true limit at which, in the organization of animals, size acquires a real importance is that of families, that is, the groups which are essentially distinguished by their form, as if form and size were correlative as far as the structure of animals is concerned. The representatives of natural families are indeed closely similar in that respect; the extreme differences are hardly anywhere tenfold within these limits and frequently only double. A few examples selected among the most natural families will show this. Omitting mankind, on account of the objections which might be made against the idea that it embraces any original diversity, let us consider the different families of Monkeys, of Bats, of Insectivora, of Carnivora, of Rodents, of Pachyderms, of Ruminants, etc., among Birds, the Vultures, the Eagles, the Falcons, the Owls, the Swallows, the Finches, the Warblers, the Humming Birds, the Doves, the Wrens, the Ostriches, the Herons, the Plovers, the Gulls, the Ducks, the Pelicans; among Reptiles, the Crocodiles, the different families of Chelonians, of Lizards, of Snakes, the Frogs proper, the Toads, etc.; among Fishes, the Sharks and Skates, the Herrings, the Codfishes, the Cyprinnodonts, the Chætodonts, the Lophobranchii, the Ostracionts, etc.; among Insects, the Sphingoidæ or the Tineina, the Longicorns or the Coccinellina, the Bomboidæ or the Brachonidæ;

among Crustacea, the Cancroidea or the Pinnotheroidæ, the Limuloidæ or the Cypridoidæ, and the Rotifera;[72] among Worms, the Dorsibranchiata or the Naioidæ; among Mollusks, the Stromboidæ or the Buccinoidæ, the Helicinoidæ or the Limnæoidæ, the Chamacea or the Cycladoidæ; among Radiata, the Asterioidæ and the Ophiuroidæ, the Hydroids and the Discophoræ, the Astræoidæ and the Actinioidæ.

Having thus recalled some facts which go to show what are the limits within which size and structure are more directly connected,[73] it is natural to infer, that since size is such an important character of species, and extends distinctly its cycle of relationship to the families or even further, it can as little be supposed to be determined by physical agents as the structure itself with which it is so closely connected, both bearing similar relations to these agents.

Life is regulated by a quantitative element in the structure of all organized beings, which is as fixed and as precisely determined as every other feature depending more upon the quality of the organs or their parts. This shows the more distinctly the presence of a specific, immaterial principle in each kind of animals and plants. All begin their existence in the condition of ovules of a microscopic size, exhibiting in all a wonderful similarity of structure. And yet these primitive ovules, so identical at first in their physical constitution, never produce anything different from the parents; all reach respectively, through a succession of unvarying changes, the same final result, the reproduction of a new being identical with the parents. How does it then happen that, if physical agents have such a powerful influence in shaping the character of organized beings, we see no trace of it in the innumerable instances in which these ovules are discharged into the elements in which they undergo their further de-

[72] Dana, *Crustacea,* pp. 1409, 1411.

[73] These remarks about the average size of animals in relation to their structure cannot fail to meet with some objections, as it is well known that under certain circumstances man may modify the normal size of a variety of plants and of domesticated animals, and that even in their natural state occasional instances of extraordinary sizes occur. But this neither modifies the characteristic average, nor is it a case which has the least bearing upon the question of origin or even the maintenance of any species, but only upon individuals, respecting which more will be found in Sect. XVI. Moreover, it should not be overlooked that there are limits to these variations and that though animals and plants may be placed under influences conducive to a more or less voluminous growth, yet it is chiefly under the agency of man that such changes reach their extremes. (See also Sect. XV.)

velopment, at a period when the germ they contain, has not yet assumed any of those more determined characteristics which distinguish the full-grown animal or the perfect plant? Do physicists know a law of the material world which presents any such analogy to these phenomena, that it could be considered as accounting for them?

In this connection it should be further remembered that these cycles of size characteristic of different families are entirely different for animals of different types, though living together under identical circumstances.

SECTION XIV

RELATIONS BETWEEN THE SIZE OF ANIMALS AND THE MEDIUMS IN WHICH THEY LIVE

It has just been remarked, that animals of different types, even when living together, are framed in structures of different size. Yet life is so closely combined with the elements of nature, that each type shows decided relations, within its own limits, to these elements as far as size is concerned.[74] The aquatic Mammalia as a whole are larger than the terrestrial ones; so are the aquatic Birds and the aquatic Reptiles. In families which are essentially terrestrial the species which take to the water are generally larger than those which remain permanently terrestrial, as for instance, the Polar Bear, the Beaver, the Coypu, and the Capivara. Among the different families of aquatic Birds those of their representatives which are more terrestrial in their habits are generally smaller than those which live more permanently in water. The same relation is observed in the different families of Insects which number aquatic and terrestrial species. It is further remarkable that among aquatic animals the fresh-water types are inferior in size to the marine ones; the marine Turtles are all larger than the largest inhabitants of our rivers and ponds, the more aquatic Trionyx larger than the Emyds, and among these the more aquatic Chelydra larger than the true Emys, and these

[74] Isidore Geoffroy St. Hilaire, *Recherches zoologiques et physiologiques sur les variations de la taille chez les animaux et dans les races humaines* (Paris, 1831). See also my paper upon the "Natural Relations between Animals and the Elements . . . ," *American Journal of Science,* IX (2d ser., 1850), 369–394.

generally larger than the more terrestrial Clemmys or the Cistudo. The class of Fishes has its largest representatives in the sea; fresh-water fishes are on the whole dwarfs, in comparison to their marine relatives, and the largest of them, our Sturgeons and Salmons, go to the sea. The same relations obtain among Crustacea; to be satisfied of the fact we need only compare our Crawfishes with the Lobsters, our Apus with Limulus, etc. Among Worms the Earthworms and Leeches furnish a still wider range of comparisons when contrasted with the marine types. Among Gasteropods and Acephala this obtains to the same extent; the most gigantic Ampullariæ and Anodontæ are small in comparison to certain Fusus, Voluta, Tritonium, Cassis, Strombus, or to the Tridacna. Among Radiata even, which are all marine, with the exception of the single genus Hydra, this rule holds good, as the fresh water Hydroids are among the smallest Acalephs known.

This coincidence upon such an extensive scale seems to be most favorable to the view that animals are modified by the immediate influence of the elements; yet I consider it as affording one of the most striking proofs that there is no causal connection between them. Were it otherwise, the terrestrial and the aquatic representatives of the same family could not be so similar as they are in all their essential characteristics, which actually stand in no relation whatsoever to these elements. What constitutes the Bear in the Polar Bear is not its adaptation to an aquatic mode of existence. What makes the Whales Mammalia bears no relation to the sea. What constitutes Earthworms, Leeches, and Eunice members of one class has no more connection with their habitat than the peculiarities of structure which unite Man, Monkeys, Bats, Lions, Seals, Beavers, Mice, and Whales into one class. Moreover, animals of different types living in the same element have no sort of similarity as to size. The aquatic Insects, the aquatic Mollusks fall in with the average size of their class, as well as the aquatic Reptiles and the aquatic Birds, or the aquatic Mammalia; but there is no common average for either terrestrial or aquatic animals of different classes taken together, and in this lies the evidence that organized beings are independent of the mediums in which they live, as far as their origin is concerned, though it is plain that when created they were made to suit the element in which they were placed.

To me these facts show that the phenomena of life are manifested in the physical world and not through or by it; that organized beings are made to conquer and assimilate to themselves the materials of the inorganic world; that they maintain their original characteristics, notwithstanding the unceasing action of physical agents upon them. And I confess I cannot comprehend how beings so entirely independent of these influences could be produced by them.

SECTION XV

PERMANENCY OF SPECIFIC PECULIARITIES IN ALL ORGANIZED BEINGS

It was a great step in the progress of science when it was ascertained that species have fixed characters and that they do not change in the course of time. But this fact, for which we are indebted to Cuvier,[75] has acquired a still greater importance since it has also been established that even the most extraordinary changes in the mode of existence and in the conditions under which animals may be placed have no more influence upon their essential characters than the lapse of time.

The facts bearing upon these two subjects are too well known now to require special illustration. I will, therefore, allude only to a few points, to avoid even the possibility of a misapprehension of my statements. That animals of different geological periods differ specifically, *en masse,* from those of preceding or following formations is a fact satisfactorily ascertained. Between two successive geological periods, then, changes have taken place among animals and plants. But none of those primordial forms of life, which naturalists call species, are known to have changed during any of these periods. It cannot be denied that the species of different successive periods are supposed by some naturalists to derive their distinguishing features from changes which have taken place in those of preceding ages; but this is a mere supposition, supported neither by physiological nor by geological evidence, and the assumption that animals and plants may change in a similar manner during one and the same period is equally

[75] *Recherches sur les ossemens fossiles* . . . (2d ed., 5 vols., Paris, 1821–1824), I, cxli.

gratuitous. On the contrary, it is known by the evidence funished by the Egyptian monuments and by the most careful comparison between animals found in the tombs of Egypt with living specimens of the same species obtained in the same country that there is not the shadow of a difference between them for a period of about five thousand years. These comparisons, first instituted by Cuvier, have proved that as far as it has been possible to carry back the investigation, it does not afford the beginning of an evidence that species change in the course of time, if the comparisons be limited to the same great cosmic epoch.[76] Geology only shows that at different periods[77] there

[76] [This was a standard argument of special creationists. They hardly understood that in terms of the life history of species five thousand years was an infinitesimal span of time.]

[77] I trust no reader will be so ignorant of the facts here alluded to as to infer from the use of the word "period" for different eras and epochs of great length, each of which is characterized by different animals, that the differences these animals exhibit is in itself evidence of a change in the species. The question is whether any changes take place during one or any of these periods. It is almost incredible how loosely some people will argue upon this point from a want of knowledge of the facts, even though they seem to reason logically. A distinguished physicist has recently taken up this subject of the immutability of species and called in question the logic of those who uphold it. I will put his argument into as few words as possible and show, I hope, that it does not touch the case. "Changes are observed from one geological period to another; species which do not exist at an earlier period are observed at a later period, while the former have disappeared; and though each species may have possessed its peculiarities unchanged for a lapse of time, the fact that when long periods are considered, all those of an earlier period are replaced by new ones at a later period, proves that species change in the end, provided a sufficiently long period of time is granted." I have nothing to object to the statement of facts, as far as it goes, but I maintain that the conclusion is not logical. It is true that species are limited to particular geological epochs; it is equally true that in all geological formations those of successive periods are different, one from the other. But because they so differ, does it follow that they have changed and not been exchanged for or replaced by others? The length of time taken for the operation has nothing to do with the argument. Granting myriads of years for each period, no matter how many or how few, the question remains simply this: When the change takes place, does it take place spontaneously under the action of physical agents, according to their law, or is it produced by the intervention of an agency not in that way at work before or afterwards? A comparison may explain my view more fully. Let a lover of the fine arts visit a museum arranged systematically and in which the works of the different schools are placed in chronological order; as he passes from one room to another, he beholds changes as great as those the paleontologist observes in passing from one system of rocks to another. But because these works bear a closer resemblance as they belong to one or the other school, or to periods following one another closely, would the critic be in any way justified in assuming that the earlier works have changed into those of a later period, or to deny that they are the works of artists living and active at the time of their production? The question about the immutability of species is identical with this supposed case. It is not because species have lasted for a longer or shorter time in past ages that naturalists consider them as immutable, but because in the whole series of geological ages, taking the entire lapse of time which has passed since

have existed different species; but no transition from those of a preceding into those of the following epoch has ever been noticed anywhere; and the question alluded to here is to be distinguished from that of the origin of the differences in the bulk of species belonging to two different geological eras. The question we are now examining involves only the fixity or mutability of species during one epoch, one era, one period in the history of our globe. And nothing furnishes the slightest argument in favor of their mutability; on the contrary, every modern investigation[78] has only gone to confirm the results first obtained by Cuvier and his views that species are fixed.

It is something to be able to show by monumental evidence, and by direct comparison that animals and plants have undergone no change for a period of about five thousand years.[79] This result has had the greatest influence upon the progress of science, especially with reference to the consequences to be drawn from the occurrence in the series of geological formations of organized beings as highly diversified in each epoch as those of the present day;[80] it has laid the foundation for the conviction, now universal among well informed naturalists, that this globe has been in existence for innumerable ages and that the length of time elapsed since it first became inhabited cannot be counted in years. Even the length of the period to which we belong is still a problem, notwithstanding the precision with which certain systems of chronology would fix the creation of

---

the first introduction of animals or plants upon earth, not the slightest evidence has yet been produced that species are actually transformed one into the other. We only know that they are different at different periods, as are works of art of different periods and of different schools; but as long as we have no other data to reason upon than those geology has furnished to this day, it is as unphilosophical and illogical, because such differences exist, to assume that species do change and have changed, that is, are transformed or have been transformed, as it would be to maintain that works of art change in the course of time. We do not know how organized beings have originated, it is true; no naturalist can be prepared to account for their appearance in the beginning or for their difference in different periods; but enough is known to repudiate the assumption of their transmutation, as it does not explain the facts and shuts out further attempts at proper investigations. See Powell, *Essays,* p. 412, *et seq.,* and Essay 3d, generally.

[78] Karl S. Kunth, "Recherches sur les plantes trouvées dans les tombeaux égyptiens," *Annales des sciences naturelles,* VIII (1826), 11.

[79] It is not for me to discuss the degree of reliability of the Egyptian chronology; but as far as it goes, it shows that from the oldest periods ascertained, animals have been what they are now.

[80] Agassiz, "The Primitive Diversity . . . ," *American Journal of Science,* XVII (2d ser., 1854), 309–354.

man.[81] There are, however, many circumstances which show that the animals now living have been for a much longer period inhabitants of our globe than is generally supposed. It has been possible to trace the formation and growth of our coral reefs, especially in Florida,[82] with sufficient precision to ascertain that it must take about eight thousand years for one of those coral walls to rise from its foundation to the level of the surface of the ocean. There are around the southernmost extremity of Florida alone four such reefs concentric with one another, which can be shown to have grown up one after the other. This gives for the beginning of the first of these reefs an age of over thirty thousand years; and yet the corals by which they were all built up are the same identical species in all of them. These facts, then, furnish as direct evidence as we can obtain in any branch of physical inquiry that some, at least, of the species of animals now existing have been in existence over thirty thousand years[83] and have not undergone the slightest change during the whole of that period.[84] And yet these four concentric reefs are only the most distinct of that region; others, less extensively investigated thus far, lie to the northward; indeed, the whole peninsula of Florida consists altogether of coral reefs annexed to one another in the course of time and containing only fragments of corals and shells, etc., identical with those now living upon that coast.[85] Now, if a width of five miles is a fair average for one coral reef growing under the circumstances under which the concentric reefs of Florida are seen now to follow one another,

[81] Nott and Gliddon, *Types of Mankind,* p. 653.

[82] See "Extract from the Report of Professor Agassiz . . . ," United States Coast Survey, *Annual Report . . . 1851* (Washington, 1852), pp. 145–160. A renewed examination of the reefs of Florida has satisfied me that this estimate falls short of the reality by a great deal. The rate of growth of the corals, ascertained by direct observation, is not half so rapid as I had been led to assume at first.

[83] I am now satisfied that the age of this reef is not overstated if estimated at one hundred thousand years, so slow are the operations of nature.

[84] Those who feel inclined to ascribe the differences which exist between species of different geological periods to the modifying influence of physical agents, and who look to the changes now going on among the living for the support of such an opinion, and may not be satisfied that the facts just mentioned are sufficient to prove the immutability of species, but may still believe that a longer period of time would yet do what thirty thousand years have not done, I beg leave to refer, for further consideration, to the charming song of Ludovici de Chamisso, entitled *Tragishe Geschichte,* and beginning as follows: " 's war Einer dem's zu Herzen ging."

[85] [Agassiz's complete account of his Florida researches was published posthumously as "Report on the Florida Reefs . . . ," *Bulletin,* Museum of Comparative Zoology, VII (1882), whole no.]

and this regular succession should extend only as far north as Lake Ogeechobee, for two degrees of latitude, this would give about two hundred thousand years for the period of time which was necessary for that part of the peninsula of Florida which lies south of Lake Ogeechobee to rise to its present southern extent above the level of the sea, and during which no changes have taken place in the character of the animals of the Gulf of Mexico.[86]

It is very prejudicial to the best interests of science to confound questions that are entirely different, merely for the sake of supporting a theory; yet this is constantly done, whenever the question of the fixity of species is alluded to. A few more words upon this point will, therefore, not be out of place here.

I will not enter into a discussion upon the question whether any species is found identically the same in two successive formations, as I have already examined it at full length elsewhere,[87] and it may be settled finally one way or the other, without affecting the proposition now under consideration; for it is plain, that if such identity could be proved, it would only show more satisfactorily how tenacious species are in their character to continue to live through all the physical changes which have taken place between two successive geological periods. Again, such identity once proved would leave it still doubtful whether their representatives in two successive epochs are descendants one of the other, as we have already strong evidence in favor of the separate origin of the representatives of the same species in separate geographical areas.[88] The case of closely allied but different species occurring in successive periods, yet limited respectively in their epochs, affords in the course of time a parallel to the case of closely allied, so-called, representative species occupying different areas in space, which no sound naturalist would suppose now to be derived one from the other. There is no more reason to suppose equally allied species following one another in time to be derived one from the other; and all that has been said in preceding paragraphs respecting the differences observed between species occurring

[86] According to facts recently observed and alluded to above, double that time at least has elapsed since their first appearance in these waters.

[87] Agassiz, *Études critiques sur les Mollusques fossiles* (4to., Neuchâtel, 1831–1845); *Monographies d'Echinodermes vivans et fossiles* (4 nos., 4to., Neuchâtel, 1838–1842); *Recherches sur les Poissons fossiles* (5 vols., Neuchâtel, 1833–1843).

[88] See Sect. x, where the case of representative species is considered.

in different geographical areas, applies with the same force to species succeeding each other in the course of time.

When domesticated animals and cultivated plants are mentioned as furnishing evidence of the mutability of species the circumstance is constantly overlooked or passed over in silence that the first point to be established respecting them, in order to justify any inference from them against the fixity of species, would be to show that each of them has originated from one common stock, which, far from being the case, is flatly contradicted by the positive knowledge we have that the varieties of several of them at least are owing to the entire amalgamation of different species.[89] The Egyptian monuments show further that many of those so-called varieties which are supposed to be the product of time are as old as any other animals which have been known to man; at all events, we have no tradition, no monumental evidence of the existence of any wild animal older than that which represents domesticated animals, already as different among themselves as they are now.[90] It is therefore quite possible that the different races of domesticated animals were originally distinct species, more or less mixed now, as the different races of men are. Moreover, neither domesticated animals nor cultivated plants, nor the races of men, are the proper subjects for an investigation respecting the fixity or mutability of species, as all involve already the question at issue in the premises which are assumed in introducing them as evidence in the case. With reference to the different breeds of our domesticated animals, which are known to be produced by the management of man, as well as certain varieties of our cultivated plants, they must be well distinguished from permanent races, which, for aught we know, may be primordial; for breeds are the result of the fostering care of man; they are the product of the limited influence and control the human mind has over organized beings, and not the free product of mere physical agents. They show therefore that even the least important changes which may take place during one and the same cosmic period among animals and plants are controlled by an intellectual power and do not result from the immediate action of physical causes.

So far then from disclosing the effects of physical agents, whatever

[89] Our fowls, for instance.

[90] Nott and Gliddon, *Types of Mankind,* p. 386.

changes are known to take place in the course of time among organized beings appear as the result of an intellectual power, and go therefore to substantiate the view that all the differences observed among finite beings are ordained by the action of the Supreme Intellect, and not determined by physical causes. This position is still more strengthened when we consider that the differences which exist between different races of domesticated animals and the varieties of our cultivated plants, as well as among the races of men, are permanent under the most diversified climatic influences; a fact which the extensive migrations of the civilized nations daily proves more extensively, and which stands in direct contradiction to the supposition that such or similar influences could have produced them.

When considering the subject of domestication in particular, it ought further to be remembered that every race of men has its own peculiar kinds of domesticated animals and of cultivated plants which exhibit much fewer varieties among them, in proportion as those races of men have had little or no intercourse with other races, than the domesticated animals of those nations which have been formed by the mixture of several tribes.

It is often stated that the ancient philosophers have solved satisfactorily all the great questions interesting to man, and that modern investigations, though they have grasped with new vigor and illuminated with new light all the phenomena of the material world, have added little or nothing in the field of intellectual progress. Is this true? There is no question so deeply interesting to man as that of his own origin and the origin of all things. And yet antiquity had no knowledge concerning it; things were formerly believed either to be from eternity or to have been created at one time. Modern science, however, can show in the most satisfactory manner that all finite beings have made their appearance successively and at long intervals, and that each kind of organized beings has existed for a definite period of time in past ages, and that those now living are of comparatively recent origin. At the same time, the order of their succession and their immutability during such cosmic periods show no causal connection with physical agents and the known sphere of action of these agents in nature, but argue in favor of repeated interventions on the part of the Creator. It seems really surprising that while such an intervention is admitted by all except the strict materialists, for the establishment of the laws regulating the inorganic world, it is yet denied by so many physicists with reference to the introduction of organized beings at different successive periods. Does this not rather go to show the imperfect acquaintance of these investigators with the conditions under which life is manifested and with the essential difference there is between the phenomena of the organic and those of the physical world, than to furnish any evidence that the organic world is the product of physical causes?

SECTION XVI

RELATIONS BETWEEN ANIMALS AND PLANTS AND THE SURROUNDING WORLD[91]

Every animal and plant stands in certain definite relations to the surrounding world, some, however, like the domestic animals and cultivated plants, being capable of adapting themselves to various conditions more readily than others; but even this pliability is a characteristic feature. These relations are highly important in a systematic point of view and deserve the most careful attention on the part of naturalists. Yet the direction zoological studies have taken since comparative anatomy and embryology began to absorb almost entirely the attention of naturalists has been very unfavorable to the investigation of the habits of animals, in which their relations to one another and to the conditions under which they live are more especially exhibited. We have to go back to the authors of the preceding century[92] for the most interesting accounts of the habits of animals, as among modern writers there are few who have devoted their chief attention to this subject.[93] So little, indeed, is its importance now appreciated, that the students of this branch of natural history are hardly

[91] [This section is an example of Agassiz's superior ability to identify basic questions and to make incisive observations in zoology. It is still pertinent from a modern viewpoint.]

[92] René Antoine de Réaumur, *Mémoires pour servir à l'histoire des Insectes* (6 vols., Paris, 1734–1742); Buffon, *Histoire naturelle générale et particulière* (44 vols., Paris, 1749–1804).

[93] John J. Audubon, *Ornithological Biography, or an Account of the Habits of the Birds of the United States of America* (5 vols., Edinburgh and Philadelphia, 1831–1839); Thaddeus W. Harris, *Report on the Insects of Massachusetts Injurious to Vegetation* (Cambridge, Mass., 1841; 2d ed., Boston, 1852).

acknowledged as peers by their fellow investigators, the anatomists and physiologists, or the systematic zoologists. And yet, without a thorough knowledge of the habits of animals, it will never be possible to ascertain with any degree of precision the true limits of all those species which descriptive zoologists have of late admitted with so much confidence into their works. And after all, what does it matter to science, that thousands of species more or less should be described and entered in our systems, if we know nothing about them? A very common defect of the works relating to the habits of animals has no doubt contributed to detract from their value and to turn the attention in other directions: their purely anecdotic character or the circumstance that they are too frequently made the occasion for narrating personal adventures. Nevertheless, the importance of this kind of investigation can hardly be overrated; and it would be highly desirable that naturalists should turn again their attention that way, now that Comparative Anatomy and Physiology, as well as Embryology, may suggest so many new topics of inquiry and the progress of Physical Geography has laid such a broad foundation for researches of this kind. Then we may learn with more precision how far the species described from isolated specimens are founded in nature, or how far they may be only a particular stage of growth of other species; then we shall know what is yet too little noticed, how extensive the range of variations is among animals observed in their wild state, or rather how much individuality there is in each and all living beings. So marked, indeed, is this individuality in many families, — and that of Turtles affords a striking example of this kind, — that correct descriptions of species can hardly be drawn from isolated specimens, as is constantly attempted to be done. I have seen hundreds of specimens of some of our Chelonians, among which there were not two identical. And truly the limits of this variability constitutes one of the most important characters of many species; and without precise information upon this point for every genus it will never be possible to have a solid basis for the distinction of species. Some of the most perplexing questions in Zoology and Palæontology might long ago have been settled had we had more precise information upon this point, and were it better known how unequal in this respect different groups of the animal kingdom are when compared with one another. While the individuals of some species seem all different and might be described as different species if seen isolated or obtained from different regions, those of other species appear all as cast in one and the same mould. It must be therefore at once obvious how different the results of the comparison of one fauna with another may be, if the species of one have been studied accurately for a long period, by resident naturalists, and the other is known only from specimens collected by chance travelers; or if the fossil representatives of one period are compared with living animals, without both faunæ having first been revised according to the same standard.[94]

Another deficiency in most works relating to the habits of animals consists in the absence of general views and of comparisons. We do not learn from them how far animals related by their structure are similar in their habits, and how far these habits are the expression of their structure. Every species is described as if it stood alone in the world; its peculiarities are mostly exaggerated, as if to contrast more forcibly with all others. Yet how interesting would be a comparative study of the mode of life of closely allied species; how instructive a picture might be drawn of the resemblance there is in this respect between species of the same genus and of the same family. The more I learn upon this subject, the more am I struck with the similarity in the very movements, the general habits, and even in the intonation of the voices of animals belonging to the same family; that is to say, between animals agreeing in the main in form, size, structure, and mode of development. A minute study of these habits, of these movements, of the voice of animals cannot fail therefore to throw additional light upon their natural affinities.

While I thus acknowledge the great importance of such investigations with reference to the systematic arrangement of animals, I cannot help regretting deeply that they are not more highly valued with reference to the information they might secure respecting the animals themselves, independently of any system. How much is there

[94] In this respect I would remark that most of the cases in which specific identity has been affirmed between living and fossil species, or between the fossils of different geological periods, belong to families which present either great similarity or extraordinary variability, and in which the limits of species are therefore very difficult to establish. Such cases should be altogether rejected in the investigation of general questions involving fundamental principles, as are untrustworthy observations always in other departments of science. Compare further my paper upon the primitive diversity and number of animals, quoted above, in which this point is specially considered.

not left to study with respect to every species after it is named and classified. No one can read Naumann's Natural History of the German Birds[95] without feeling that natural history would be much further advanced if the habits of all other animals had been as accurately investigated and as minutely recorded; and yet that work contains hardly anything of importance with reference to the systematic arrangement of birds. We scarcely possess the most elementary information necessary to discuss upon a scientific basis the question of the instincts, and in general the faculties of animals, and to compare them together and with those of man, not only because so few animals have been thoroughly investigated, but because so much fewer still have been watched during their earlier periods of life, when their faculties are first developing; and yet how attractive and instructive this growing age is in every living being! Who, for instance, could believe for a moment longer that the habits of animals are in any degree determined by the circumstances under which they live, after having seen a little turtle of the genus Chelydra, still enclosed in its egg-shell, which it hardly fills half-way, with a yolk bag as large as itself hanging from its lower surface and enveloped in its amnios and in its allantois, with the eyes shut, snapping as fiercely as if it could bite without killing itself? Who can watch the Sunfish (*Pomotis vulgaris*) hovering over its eggs and protecting them for weeks, or the Catfish (*Pimelodus Catus*) move about with its young, like a hen with her brood, without remaining satisfied that the feeling which prompts them in these acts is of the same kind as that which attaches the Cow to her suckling, or the child to its mother? Is there an investigator, who having once recognized such a similarity between certain faculties of Man and those of the higher animals, can feel prepared in the present stage of our knowledge to trace the limit where this community of nature ceases? And yet to ascertain the character of all these faculties there is but one road, the study of the habits of animals and a comparison between them and the earlier stages of development of Man. I confess I could not say in what the mental faculties of a child differ from those of a young Chimpanzee.

Now that we have physical maps of almost every part of the

[95] [Johann Andreas Naumann, *Naturgeschichte der Vögel Deutschlands, nach eignen Erfahrungen entworfen. Durchaus umgearbeitet . . . von dessen Sohne Johann Friedrich Naumann,* 12 vols., Leipzig, 1822–1844].

globe,[96] exhibiting the average temperature of the whole year and of every season upon land and sea; now that the average elevation of the continents above the sea and that of the most characteristic parts of their surface, their valleys, their plains, their table-lands, their mountain systems — are satisfactorily known; now that the distribution of moisture in the atmosphere, the limits of the river systems, the prevailing direction of the winds, the course of the currents of the ocean, are not only investigated, but mapped down, even in school atlases; now that the geological structure of nearly all parts of the globe has been determined with tolerable precision — zoologists have the widest field and the most accurate basis to ascertain all the relations which exist between animals and the world in which they live.

Having thus considered the physical agents with reference to the share they may have had in calling organized beings into existence and satisfied ourselves that they are not the cause of their origin, it now remains for us to examine more particularly these relations as an established fact, as conditions in which animals and plants are placed at the time of their creation, within definite limits of action and reaction between them; for, though not produced by the influence of the physical world, organized beings live in it, they are born in it, they grow up in it, they multiply in it, they assimilate it to themselves or feed upon it, they have even a modifying influence upon it within the same limits, as the physical world is subservient to every manifestation of their life. It cannot fail, therefore, to be highly interesting and instructive to trace these connections, even without any reference to the manner in which they were established, and this is the proper sphere of investigation in the study of the habits of animals. The behavior of each kind toward its fellow-beings and with reference to the conditions of existence in which it is placed constitutes a field of inquiry of the deepest interest, as extensive as it is complicated. When properly investigated, especially within the sphere which constitutes more particularly the essential characteristics of each species of animals and plants, it is likely to afford the most direct evidence of the unexpected independence of physical influences of organized beings, if I mistake not the evidence I have

[96] Alexander Keith Johnston, *The Physical Atlas of Natural Phenomena* (Edinburgh, 1848).

myself been able to collect. What can there be more characteristic of different species of animals than their motions, their plays, their affections, their sexual relations, their care of their young, the dependence of these upon their parents, their instincts, etc., etc.; and yet there is nothing in all this which depends in the slightest degree upon the nature or the influence of the physical conditions in which they live. Even their organic functions are independent of these conditions to a degree unsuspected, though this is the sphere of their existence which exhibits the closest connections with the world around.

Functions have so long been considered as the test of the character of organs, that it has almost become an axiom in Comparative Anatomy and Physiology that identical functions presuppose identical organs. Most of our general works upon Comparative Anatomy are divided into chapters according to this view. And yet there never was a more incorrect principle, leading to more injurious consequences, more generally adopted. That naturalists should not long ago have repudiated it is the more surprising, as every one must have felt again and again how unsound it is. The organs of respiration and circulation of fishes afford a striking example. How long have not their gills been considered as the equivalent of the lungs of the higher Vertebrata, merely because they are breathing organs; and yet these gills are formed in a very different way from the lungs; they bear very different relations to the vascular system; and it is now known that they may exist simultaneously with lungs, as in some full-grown Batrachians, and, in the earlier embryonic stages of development in all Vertebrata. There can now no longer be any doubt that they are essentially different organs and that their functions afford no test of their nature and cannot constitute an argument in favor of their organic identity. The same may be said of the vascular system of the fishes. Cuvier[97] described their heart as representing the right auricle and the right ventricle, because it propels the blood it contains to the gills, in the same manner as the right ventricle propels the blood to the lungs of the warm-blooded animals; yet Embryology has taught us that such a comparison based upon the special relations of the heart of fishes is unjustifiable. The air sacs of certain spiders have also been considered as lungs because they per-

[97] Cuvier, *Règne animal* (2d ed.), II, 122.

form similar respiratory functions, and yet they are only modified tracheæ which are constructed upon such a peculiar plan and stand in such different relations to the peculiar kind of blood of the Articulata, that no homology can be traced between them and the lungs of Vertebrata, no more than between the so-called lungs of the air-breathing Mollusks, whose aërial respiratory cavity is only a modification of the peculiar kind of gills observed in other Mollusks. Examples might easily be multiplied; I will, however, only allude further to the alimentary canal of Insects and Crustacea, with its glandular appendages, formed in such a different way from that of Vertebrata, or Mollusks, or Radiata, to their legs and wings, etc., etc. I might allude also to what has been called the foot in Mollusks, did it not appear like pretending to suppose that anyone entertains still an idea that such a name implies any similarity between their locomotive apparatus and that of Vertebrata or Articulata; and yet, the very use of such a name misleads the student, and even some of the coryphees of our science have not freed themselves of such and similar extravagant comparisons, especially with reference to the solid parts of the frame of the lower animals.

The identification of functions and organs was a natural consequence of the prevailing ideas respecting the influence physical agents were supposed to have upon organized beings. But as soon as it is understood how different the organs may be which in animals perform the same function, organization is at once brought into such a position to physical agents as makes it utterly impossible to maintain any genetic connection between them. A fish, a crab, a mussel, living in the same waters, breathing at the same source, should have the same respiratory organs, if the elements in which these animals live had anything to do with shaping their organization. I suppose no one can be so short-sighted, as to assume that the same physical agents acting upon animals of different types must produce in each peculiar organs, and not to perceive that such an assumption implies the very existence of these animals, independently of the physical agents. But this mistake recurs so constantly in discussions upon this and similar topics, that, trivial as it is, it requires to be rebuked.[98]

[98] I hope the day is not far distant when zoologists and botanists will equally disclaim having shared in the physical doctrines more or less prevailing now, respecting the origin and existence of organized beings. Should the time come when my present

On the contrary, when acknowledging an intellectual conception, as the preliminary step in the existence not only of all organized beings, but of everything in nature, how natural to find that while diversity is introduced in the plan, in the complication and the details of structure of animals, their relations to the surrounding media are equally diversified, and consequently the same functions may be performed by the most different apparatus!

SECTION XVII

RELATIONS OF INDIVIDUALS TO ONE ANOTHER

The relations in which individuals of the same species of animals stand to one another are not less determined and fixed than the relations of species to the surrounding elements, which we have thus far considered. The relations which individual animals bear to one another are of such a character that they ought long ago to have been considered as proof sufficient that no organized being could ever have been called into existence by another agency than the direct intervention of a reflective mind. It is in a measure conceivable that physical agents might produce something like the body of the lowest kinds of animals or plants, and that under identical circumstances the same thing may have been produced again and again, by the repetition of the same process; but that upon closer analysis of the possibilities of the case it should not have at once appeared how incongruous the further supposition is, that such agencies could delegate the power of reproducing what they had just called into existence, to those very beings, with such limitations that they could never reproduce anything but themselves, I am at a loss to understand. It will no more do to suppose that from simpler structures such a process may end in the production of the most perfect, as every step implies an addition of possibilities not even included in the original case. Such a delegation of power can only be an act of

efforts may appear like fighting against windmills, I shall not regret having spent so much labor in urging my fellow-laborers in a right direction; but at the same time I must protest now and forever against the bigotry spreading in some quarters, which would press upon science doctrines not immediately flowing from scientific premises and check its free progress.

intelligence; while between the production of an indefinite number of organized beings, as the result of a physical law, and the reproduction of these same organized beings by themselves, there is no necessary connection. The successive generations of any animal or plant cannot stand, as far as their origin is concerned, in any causal relation to physical agents, if these agents have not the power of delegating their own action to the full extent to which they have already been productive in the first appearance of these beings; for it is a physical law that the resultant is equal to the forces applied. If any new being has ever been produced by such agencies, how could the successive generations enter, at the time of their birth, into the same relations to these agents as their ancestors, if these beings had not in themselves the faculty of sustaining their character, in spite of these agents? Why, again, should animals and plants at once begin to decompose under the very influence of all those agents which have been subservient to the maintenance of their life as soon as life ceases, if life is limited or determined by them?

There exist between individuals of the same species relations far more complicated than those already alluded to, which go still further to disprove any possibility of causal dependence of organized beings upon physical agents. The relations upon which the maintenance of species is based throughout the animal kingdom, in the universal antagonism of sex and the infinite diversity of these connections in different types, have really nothing to do with external conditions of existence; they indicate only relations of individuals to individuals, beyond their connections with the material world in which they live. How, then, could these relations be the result of physical causes, when physical agents are known to have a specific sphere of action in no way bearing upon this sphere of phenomena?

For the most part, the relations of individuals to individuals are unquestionably of an organic nature, and as such have to be viewed in the same light as any other structural feature; but there is much also in these connections that partakes of a psychological character, taking this expression in the widest sense of the word.

When animals fight with one another, when they associate for a common purpose, when they warn one another in danger, when they come to the rescue of one another, when they display pain or joy, they manifest impulses of the same kind as are considered among the moral attributes of man. The range of their passions is even as extensive as that of the human mind, and I am at a loss to perceive a difference of kind between them, however much they may differ in degree and in the manner in which they are expressed. The gradations of the moral faculties among the higher animals and man are, moreover, so imperceptible, that to deny to the first a certain sense of responsibility and consciousness would certainly be an exaggeration of the difference between animals and man. There exists, besides, as much individuality, within their respective capabilities, among animals as among men, as every sportsman, or every keeper of menageries, or every farmer and shepherd can testify who has had a large experience with wild or tamed or domesticated animals.

This argues strongly in favor of the existence in every animal of an immaterial principle similar to that which, by its excellence and superior endowments, places man so much above animals.[99] Yet the

[99] It might easily be shown that the exaggerated views generally entertained of the difference existing between man and monkeys are traceable to the ignorance of the ancients, and especially the Greeks, to whom we owe chiefly our intellectual culture, of the existence of the Orang-Outang and the Chimpanzee. The animals most closely allied to man known to them were the Red Monkey, the Baboon, and the Barbary Ape. A modern translation of Aristotle, it is true, makes him say that monkeys form the transition between man and quadrupeds: (Aristoteles, *Naturgeschichte der Thiere, von Dr. Franz Strack,* Frankfurt am Main, 1816, p. 65) but the original says no such thing. In the *History of Animals,* Book II, Chap. 8. we read only ἔνια δὲ τῶν ζώων ἐπαμφοτερίζει τὴν φύσιν τῷ τε ἀνθρώπῳ καὶ τοῖς τετράποσιν. ["Some animals share the properties of man and the quadrupeds, as the ape, the monkey, and the Baboon." D'Arcy Wentworth Thompson, tr., *Historia Animalium* (Oxford, 1910), II. 8. 502a16–18.] There is a wide difference between "partaking of the nature of both man and the quadrupeds" and "forming a transition between man and the quadrupeds." The whole chapter goes on enumerating the structural similarity of the three monkeys named above with man, but the idea of a close affinity is not even expressed, and still less that of a transition between man and the quadrupeds. The writer, on the contrary, dwells very fully upon the marked differences they exhibit and knows as well as any modern anatomist has ever known that monkeys have four hands. ἔχει δὲ καὶ βραχίονας, ὥσπερ ἄνθρωπος . . . ἰδίους δὲ τοὺς πόδας· εἰσὶ γὰρ οἷον χεῖρες μεγάλαι. Καὶ οἱ δάκτυλοι ὥσπερ οἱ τῶν χειρῶν, ὁ μέγας μακρότατος· καὶ τὸ κάτω τοῦ ποδὸς χειρὶ ὅμοιον, πλὴν ἐπὶ τὸ μῆκος τὸ τῆς χειρὸς ἐπὶ τα ἔσχατα τεῖνον καθάπερ θέναρ. Τοῦτο δὲ ἐπ' ἄκρου σκληρότερον, κακῶς και ἀμυδρῶς μιμούμενον πτέρνην. ["The ape . . . like man . . . its feet are exceptional in kind. That is, they are like large hands, and the toes are like fingers, with the middle one the longest of all, and the under part of the foot is like a hand except for its length, and stretches out towards the extremities like the palm of the hand, and this palm at the after end is unusually hard, and in a clumsy kind of way resembles a heel." Thompson, tr., *Hist. Anim.*, II. 8. 502b2.]

It is strange that these clear and precise distinctions should have been so entirely forgotten in the days of Linnæus that the great reformer in Natural History had to confess in the year 1746 that he knew no character by which to distinguish man form monkeys. *Fauna Suecica. . . .* (Stockholm), Praefat., p. 2. "Nullum characterem adhuc eruere potui, unde home a simia internoscatur." But it is not upon structural simi-

principle exists unquestionably, and whether it be called soul, reason, or instinct, it presents in the whole range of organized beings a series of phenomena closely linked together; and upon it are based not only the higher manifestations of the mind, but the very permanence of the specific differences which characterize every organism. Most of the arguments of philosophy in favor of the immortality of man apply equally to the permanency of this principle in other living beings. May I not add that a future life, in which man should

larity or difference alone that the relations between man and animals have to be considered. The psychological history of animals shows that as man is related to animals by the plan of his structure, so are these related to him by the character of those very faculties which are so transcendent in man as to point at first to the necessity of disclaiming for him completely any relationship with the animal kingdom. Yet the natural history of animals is by no means completed after the somatic side of their nature has been thoroughly investigated; they, too, have a psychological individuality, which, though less fully studied, is nevertheless the connecting link between them and man. I cannot therefore agree with those authors who would disconnect mankind from the animal kingdom and establish a distinct kingdom for man alone, as Ehrenberg (*Das Naturreich des Menschen,* 1835) and lately Isidore Geoffroy St. Hilaire, (*Histoire naturelle générale des règnes organiques,* 3 vols., Paris, 1854–1862, I, pt. 2, 167) have done. Compare, also, Chap. II, where it is shown for every kind of group of the animal kingdom that the amount of their difference one from the other never affords a sufficient ground for removing any of them into another category. A close study of the dog might satisfy every one of the similarity of his impulses with those of man, and those impulses are regulated in a manner which discloses psychical faculties in every respect of the same kind as those of man; moreover, he expresses by his voice his emotions and his feelings, with a precision which may be as intelligible to man as the articulated speech of his fellow men. His memory is so retentive that it frequently baffles that of man. And though all these faculties do not make a philosopher of him, they certainly place him in that respect upon a level with a considerable proportion of poor humanity. The intelligibility of the voice of animals to one another and all their actions connected with such calls are also a strong argument of their perceptive power and of their ability to act spontaneously and with logical sequence in accordance with these perceptions. There is a vast field open for investigation in the relations between the voice and the actions of animals, and a still more interesting subject of inquiry in the relationship between the cycle of intonations which different species of animals of the same family are capable of uttering, which, as far as I have as yet been able to trace them, stand to one another in the same relations as the different so-called families of languages in the human family. All the *Canina* bark; the howling of the wolves, the barking of the dogs and foxes, are only different modes of barking, comparable to one another in the same relation as the monosyllabic, the agglutinating, and the inflecting languages. The *Felidæ* mew: the roaring of the lion is only another form of the mewing of our cats and the other species of the family. The *Equina* neigh or bray: the horse, the donkey, the zebra, the dauw, do not differ much in the scale of their sounds. Our cattle and the different kinds of wild bulls have a similar affinity in their intonations; their lowing differs not in kind, but only in the mode of utterance. Among birds this is perhaps still more striking. Who does not distinguish the note of any and every thrush, or of the warblers, the ducks, the fowls, etc., however numerous their species may be, and who can fail to perceive the affinity of their voices? And does this not indicate a similarity also in their mental faculties?

be deprived of that great source of enjoyment and intellectual and moral improvement which result from the contemplation of the harmonies of an organic world, would involve a lamentable loss, and may we not look to a spiritual concert of the combined worlds and all their inhabitants in presence of their Creator as the highest conception of paradise?

SECTION XVIII

METAMORPHOSES OF ANIMALS

The study of Embryology is of very recent date; the naturalists of the past century, instead of investigating the phenomena accompanying the first formation and growth of animals, were satisfied with vague theories upon reproduction.[100] It is true the metamorphoses of Insects became very early the subject of most remarkable observations, but so little was it then known that all animals undergo great changes from the first to the last stages of their growth, that metamorphosis was considered a distinguishing character of Insects. The differences between Insects in that respect are, however, already so great, that a distinction was introduced between those which undergo a complete metamorphosis, that is to say, which appear in three successive different forms, as larvæ, pupæ, and perfect insects, and those with an incomplete metamorphosis, or whose larvæ differ little from the perfect insect. The range of these changes is yet so limited in some insects, that it is not only not greater, but is even much smaller than in many representatives of other classes. We may therefore well apply the term metamorphosis to designate all the changes which animals undergo in direct and immediate succession[101] during their growth, whether these changes are great or small, provided they are correctly qualified for each type.

The study of Embryology, at first limited to the investigation of

[100] Buffon, *Discours sur la nature des animaux* (Geneva, 1754); also in his *Oeuvres complètes* (36 vols., Paris, 1774–1804).

[101] I say purposely "in direct and immediate succession," as the phenomena of alternate generation are not included in metamorphosis, and consist chiefly in the production of new germs, which have their own metamorphosis; while metamorphosis proper relates only to the successive changes of one and the same germ.

the changes which the chicken undergoes in the egg, has gradually extended to every type of the animal kingdom; and so diligent and thorough has been the study, that the first author who ventured upon an extensive illustration of the whole field, C. E. von Baer, has already presented the subject in such a clear manner, and drawn general conclusions so accurate and so comprehensive, that all subsequent researches in this department of our science, may be considered only as a further development of the facts first noticed by him and of the results he has already deduced from them.[102] It was he who laid the foundation for the most extensive generalizations respecting the mode of formation of animals; for he first discovered in 1827 the ovarian egg of Mammalia and thus showed for the first time that there is no essential difference in the mode of reproduction of the so-called viviparous and oviparous animals, and that man himself is developed in the same manner as the animals. The universal presence of eggs in all animals and the unity of their structure, which was soon afterwards fully ascertained, constitute, in my opinion, the greatest discovery in the natural sciences of modern times.[103]

It was, indeed, a gigantic step to demonstrate such an identity in the material basis of the development of all animals, when their anatomical structure was already known to exhibit such radically different plans in their full-grown state. From that time a more and more extensive investigation of the manner in which the first germ is formed in these eggs and the embryo develops itself; how its organs grow gradually out of a homogeneous mass; what changes, what complications, what connections, what functions they exhibit at every stage; how in the end the young animal assumes its final form and

[102] Without referring to the works of older writers, such as Regnier de Graaf, Marcello Malpighi, Albrecht von Haller, Caspar F. Wolff, Johann F. Meckel, Friedrich Tiedemann, etc., which are all enumerated with many others in Theodor Bischoff, "Entwickelungsgeschichte," in Rudolf Wagner, *Handwörterbuch der Physiologie* . . . (12 vols., Brunswick, 1842–1845), I, 860. I shall mention hereafter chiefly those published since. Under the influence of Ignatius Döllinger this branch of science has assumed a new character. See Carl Ernst von Baer, *Ueber Entwickelungsgeschichte der Thiere* (2 vols., Königsberg, 1828–1837), the most important work yet published. The preface is a model of candor and truthfulness and sets the merits of Döllinger in a true and beautiful light.

[103] Von Baer, *De Ovi Mammalium et Hominis Genesi* (Königsberg, 1827); Jan E. Purkinje, *Symbolæ ad ovi avium historiam ante incubationem* (Leipzig, 1830); Wagner, *Prodromus Historiæ generationis Hominis atque Animalium* . . . (Leipzig, 1836); and *Icones physiologicæ* (Leipzig, 1839).

structure and becomes a new, independent being, could not fail to be the most interesting subject of inquiry. To ascertain all this in as many animals as possible, belonging to the most different types of the animal kingdom, became soon the principal aim of all embryological investigations; and it can truly be said that few sciences have advanced with such astonishing rapidity and led to more satisfactory results.

For the actual phases of the mode of development of the different types of the animal kingdom I must refer to the special works upon this subject,[104] no general treatise embracing the most recent investigations having as yet been published; and I must take it for granted that before forming a definite opinion upon the comparisons instituted hereafter between the growth of animals and the structural gradation among full-grown animals, or the order of succession of the fossils characteristic of different geological periods, the necessary information respecting these changes has been gathered by my readers and sufficiently mastered to enable them to deal with it freely.

The embryology of Polypi has been very little studied thus far; what we know of the embryonic growth of these animals relates chiefly to the family of Actinoids. When the young is hatched, it has the form of a little club-shaped or pear-shaped body, which soon assumes the appearance of the adult, from which it differs only by having few tentacles. The mode of ramification and the multiplication by buds have, however, been carefully and minutely studied in all the families of this class. Acalephs present phenomena so peculiar that they are discussed hereafter in a special section. Their

[104] The limited attention thus far paid in this country to the study of Embryology has induced me to enumerate the works relating to this branch of science more fully than any others, in the hope of stimulating investigations in this direction. There exist upon this continent a number of types of animals, the embryological illustration of which would add immensely to the stock of our science; such are the Opossum, the Ichthyoid Batrachians, the Leipidosteus, the Amia, etc., not to speak of the opportunities which thousands of miles of sea-coast, everywhere easily accessible, afford for embryological investigations, from the borders of the Arctics to the Tropics. In connection with Embryology the question of Individuality comes up naturally. See upon this subject: Rudolf Leuckart, *Ueber den Polymorphismus der Individuen oder die Erscheinung der Arbeitscheilung in der Natur* (Giessen, 1851); Thomas Henry Huxley, "Upon Animal Individuality," *Annals and Magazine of Natural History,* IX (2d ser., 1852), 507; Edward Forbes, "On the supposed Analogy between the Life of an Individual and the Duration of a Species," *ibid.,* X (2d ser., 1852), 59; Alexander Braun, *Das Individuum der Pflanze* (Berlin, 1853), and *Betrachtungen über die Erscheinung der Verjüngung in der Natur* (Freiburg, 1849).

young are either polyplike or resemble more immediately the type of their class. Few multiply in a direct, progressive development. As to Echinoderms, they have for a long time almost entirely escaped the attention of Embryologists, but lately J. Müller[105] has published a series of most important investigations upon this class, disclosing a wonderful diversity in the mode of their development, not only in the different orders of the class, but even in different genera of the same family. The larvæ of many have a close resemblance to diminutive Ctenophoræ and may be homologized with this type of Acalephs.

As I shall hereafter refer frequently to the leading divisions of the animal kingdom, I ought to state here that I do not adopt some of the changes which have been proposed lately in the limitation of the classes and which seem to have been pretty generally received with favor. The undivided type of Radiata appears to me as one of the most natural branches of the animal kingdom, and I consider its subdivision into Coelenterata and Echinodermata as an exaggeration of the anatomical differences observed between them. As far as the plan of their structure is concerned they do not differ at all, and that structure is throughout homological. In this branch I recognize only three classes, *Polypi, Acalephæ,* and *Echinodermata.* The chief differences between the two first lies in the radiating partitions of the main cavity of the Polypi supporting the reproductive organs; moreover, the digestive cavity in this class consists of an inward fold of the upper aperture of the common sac of the body, while in Acalephs there exist radiating tubes, at least in the *proles medusina,* which extend to the margin of the body where they anastomoze, and the digestive cavity is hollowed out of the gelatinous mass of the body. This is equally true of the Hydroids, the Medusæ proper, and the Ctenophoræ; but nothing of the kind is observed among Polypi. Siphonophoræ, whether their *proles medusina* becomes free or not, and Hydroids agree in having, in the *proles medusina,* simple radiating tubes, uniting into a single circular tube around the margin of the bell-shaped disk. These two groups constitute together one natural order, in contradistinction from the Covered-eyed Medusæ, whose radiating tubes ramify towards the margin and form a complicated net of anastomoses. Morphologically,

[105] [Johannes Peter Müller, 1801–1858.]

the *proles polypoidea* of the Acalephs is as completely an Acaleph as their *proles medusina,* and whether they separate or remain connected, their structural relations are everywhere the same. A comparison of Hydractinia, which is the most common and the most polymorphous Hydroid, with our common Portuguese Man-of-War (*Physalia*) may at once show the homology of their most polymorphous individuals.

The embryology of Mollusks has been very extensively investigated, and some types of this branch are among the very best known in the animal kingdom. The natural limits of the branch itself appear, however, somewhat doubtful. I hold that it must include the Bryozoa, which lead gradually through the Brachiopods and Tunicata to the ordinary Acephala, and I would add that I have satisfied myself of the propriety of uniting the Vorticellidæ with Bryozoa. On the other hand, the Cephalopods can never be separated from the Mollusks proper as a distinct branch; the partial segmentation of their yolk no more affords a ground for their separation than the total segmentation of the yolk of Mammalia would justify their separation from the other Vertebrata. Moreover, Cephalopods are in all the details of their structure homologous with the other Mollusks. The Tunicata are particularly interesting, inasmuch as the simple Ascidians have pedunculated young which exhibit the most striking resemblance to Boltenia and form at the same time a connecting link with the compound Ascidians. The development of the Lamellibranchiata seems to be very uniform, but they differ greatly as to their breeding, many laying their eggs before the germ is formed, whilst others carry them in their gills until the young are entirely formed. This is observed particularly among the Unios, some of which, however, lay their eggs very early, while others carry them for a longer or shorter time in a special pouch of the outer gill, which presents the most diversified forms in different genera of this family. Nothing is as yet known of the development of Brachiopods. The Gasteropods exhibit a much greater diversity in their development than the Lamellibranchiata. Even among the terrestrial and aquatic Pulmonata there are striking differences. Some of the Pectinibranchiata are remarkable for the curious cases in which their eggs are hatched and the young developed to an advanced state of growth. The cases of Pyrula and Strombus are among the most extraordinary of these organic nests. The embryology of Cephalopods has been masterly illustrated by Kölliker.[106]

There is still much diversity of opinion among naturalists respecting the limits of Articulata, some being inclined to separate the Arthropoda and Worms as distinct branches, while others unite them into one. I confess I cannot see the ground for a distinction. The worm-like nature of the larvæ of the majority of Arthropods and the perfect homology of these larvæ with the true Worms seem to me to show beyond the possibility of a doubt that all these animals are built upon one and the same plan and belong therefore to one branch, which contains only three classes, if the principles laid down in my second chapter are at all correct, namely, the Worms, Crustacea, and Insects. As to the Protozoa, I have little confidence in the views generally entertained respecting their nature. Having satisfied myself that Colpoda and Paramecium are the brood of Planariæ, and Opalina that of Distoma, I see no reason, why the other Infusoria, included in Ehrenberg's division Enterodela,[107] should not also be the brood of the many lower Worms, the development of which has thus far escaped our attention. Again, a comparison of the early stages of development of the Entomostraca with Rotifera might be sufficient to show, what Burmeister, Dana, and Leydig have proved in another way, that Rotifera are genuine Crustacea, and not Worms. The vegetable character of most of the Anentera has been satisfactorily illustrated. I have not yet been able to arrive at a definite result respecting the Rhizopods, though they may represent in the type of Mollusks the stage of yolk segmentation of Gasteropods. From these remarks it should be inferred that I do not consider the Protozoa as a distinct branch of the animal kingdom, nor the Infusoria as a natural class.

Taking the class of Worms, in the widest sense, it would thus embrace the Helminths, Turbellariæ, and Annulata. The embryology of these animals still requires careful study, notwithstanding the many extensive investigations to which they have been submitted; the intestinal Worms especially continue to baffle the zeal of naturalists, even now when the leading features of their development are ascertained. The Nematoids undergo a very simple development,

[106] [Rudolf Albert von Kölliker, 1817–1905.]
[107] That Vorticellidæ are Bryozoa has already been stated above.

without alternate generations, and as some are viviparous their changes can easily be traced. The Cestods and Cystici, which were long considered as separate orders of Helminths, are now known to stand in direct genetic connection with one another, the Cystici being only earlier stages of development of the Cestods. The Trematods exhibit the most complicated phenomena of alternate generations; but as no single species has thus far been traced through all the successive stages of its transformations, doubts are still entertained respecting the genetic connection of many of the forms which appear to belong to the same organic cycle. It is also still questionable, whether Gregarinæ and Psorospermia are embryonic forms or not, though the most recent investigations render it probable that they are. The development of the Annulata, as they are now circumscribed, exhibits great variety; some resemble more the Nematods, in their metamorphoses, while others, the Leeches for instance, approximate more the type of the Trematods. The Sipunculoids appear to be more closely related to the Annulata than to the Holothurioids.

The class of Crustacea, on the contrary, may be considered as one of the best known, as far as its zoological characters and embryonic growth are concerned, the only point still questioned being the relationship of the Rotifera. In their mode of development the Lernæans, the Entomostraca proper, and the Cirripeds agree as closely with one another as they differ from the higher Crustacea. This conformity is the more interesting, as the low position the Entomostraca hold in the class of Crustacea agrees strikingly with their early appearance in geological times, while the form of the adult Cirripeds and that of the Lernæans would hardly lead one to suspect their near relationship, which has, indeed, been quite overlooked until Embryology showed that their true position is among Crustacea. In the development of the higher Crustacea their superior rank is plainly exhibited, and few types show more directly a resemblance, in their early stages of development, to the lower members of their class than the Brachyura.

In the class of Insects I include Myriapods, Arachnoids, and the true Insects, as, according to the views expressed hereafter, these natural groups constitute only different degrees of complication of the same combination of organic systems, and must therefore be considered as natural orders of one and the same class. This class, though very extensively studied in a zoological and anatomical point of view, and as far as the habits of its representatives are concerned, still requires, however, much patient work, as the early embryonic development of these animals has been much less studied than their later transformations. The type of the Arachnoids embraces two groups, the Acari and the Arachnoids proper, corresponding respectively in this class to the Entomostraca and the higher Crustacea. The embryo of the Acari resembles somewhat that of the Entomostraca, whilst that of the true Spiders recalls the metamorphosis of the higher Crustacea. On the ground of the similarity of their young, some animals, formerly referred to the class of Worms, are now considered as Arachnoids; but the limits between the aquatic Mites and the Pycnogonums are not yet quite defined.

In the branch of Vertebrata all classes have been extensively studied, and as far as the principal types are concerned, the leading features of their development are satisfactorily known. Much, however, remains to be done to ascertain the minor modifications characteristic of the different families. It may even be that further investigations will greatly modify the general classification of the whole branch. The class of Fishes may require subdivision, since the development of the Plagiostoms differs greatly from that of the ordinary fishes. As it now stands in our systems, the class of Fishes is certainly the most heterogeneous among Vertebrata. The disagreement of authors as to the limits and respective value of its orders and families may be partly owing to the unnatural circumscription of the class itself.[108] As to the Reptiles, it is already certain that the Am-

[108] The peculiarities of the development of the Plagiostoms consist not so much in the few large eggs they produce, and the more intimate connection which the embryo of some of them assumes with the parent, than in the development itself, which, not withstanding the absence of amnios and an allantois, resembles closely in its early stages that of the Reptiles proper and of the Birds, especially in the formation of the vascular system, the presence of a *sinus terminalis,* etc. Again, besides the more obvious anatomical differences existing between the Plagiostoms and the bony Fishes, it should be remembered that, as in the higher Vertebrata, the ovary is separated from the oviducts in the Sharks and Skates, and the eggs are taken up by a wide fallopian tube. That the Plagiostoms can hardly be considered simply as an order in the class of Fishes, could already be inferred from the fact that they do not constitute a natural series with the other Fishes. I would, therefore, propose the name of SELACHIANS for a distinct class embracing the Sharks, Skates, and Chimæras. Recent investigations upon the Cyclostoms show them also to differ widely from the Fishes proper, and they too ought to be separated as a distinct class, for which the name of MYZONTES may be most appropriate.

phibia and Reptiles proper, so long united as one class, constitute two distinct classes. In the main, the development of the true Reptiles agrees very closely with that of the Birds, while the Amphibians resemble more the true fishes. In no class are renewed embryological investigations, extending over a variety of families, so much needed, as in that of Birds, though the general development of these animals is, perhaps, better known than that of any other type; while the class of Mammalia has found in Bischoff a most successful and thorough investigator.

Embryology has, however, a wider scope than to trace the growth of individual animals, the gradual building up of their body, the formation of their organs, and all the changes they undergo in their structure and in their form; it ought also to embrace a comparison of these forms and the successive steps of these changes between all the types of the animal kingdom, in order to furnish definite standards of their relative standing, of their affinities, of the correspondence of their organs in all their parts. Embryologists have thus far considered too exclusively the gradual transformation of the egg into a perfect animal; there remains still a wide field of investigation to ascertain the different degrees of similarity between the successive forms an animal assumes until it has completed its growth and the various forms of different kinds of full-grown animals of the same type; between the different stages of complication of their structure in general and the perfect structure of their kindred; between the successive steps in the formation of all their parts and the various degrees of perfection of the parts of other groups; between the normal course of the whole development of one type compared with that of other types, as well as between the ultimate histological differences which all exhibit within certain limits. Though important fragments have been contributed upon these different points, I know how much remains to be done, from the little I have as yet been able to gather myself by systematic research in this direction.

I satisfied myself long ago that Embryology furnishes the most trustworthy standard to determine the relative rank among animals. A careful comparison of the successive stages of development of the higher Batrachians furnishes perhaps the most striking example of the importance of such investigations. The earlier stages of the Tadpole exemplify the structure and form of those Ichthyoids which have either no legs or very imperfect legs, with and without external gills; next it assumes a shape reminding us more of the Tritons and Salamanders and ends with the structure of the Frog or Toad. A comparison between the two latter families might prove further that the Toads are higher than the Frogs, not only on account of their more terrestrial habits (see Sect. XVI), but because the embryonic web, which to some extent still unites the fingers in the Frogs, disappears entirely in the Toads, and may be also because glands are developed in their skin which do not exist in Frogs. A similar comparison of the successive changes of a new species of Comatula discovered by Prof. Holmes,[109] in the harbor of Charleston, in South Carolina, has shown me in what relation the different types of Crinoids of past ages stand to these changes, and has furnished a standard to determine their relative rank; as it cannot be doubted that the earlier stages of growth of an animal exhibit a condition of relative inferiority, when contrasted with what it grows to be after it has completed its development and before it enters upon those phases of its existence which constitute old age and certain curious retrograde metamorphoses observed among parasites.[110]

In the young Comatula there exists a stem by which the little animal is attached, either to sea weeds or to the cirrhi of the parent; the stem is at first simple and without cirrhi, supporting a globular head upon which the so-called arms are next developed and gradually completed by the appearance of branches; a few cirrhi are at the same time developed upon the stem, which increase in number until they form a wreath between the arms and the stem. At last, the crown having assumed all the characters of a diminutive Comatula, drops off, freeing itself from the stem, and the Comatula moves freely as an independent animal.

[109] [Francis S. Holmes, 1815–1892.]

[110] [Agassiz returns to this concept repeatedly, because it represents his particular understanding of "evolution" and "development." He consistently identifies change in the individual (ontogeny) with change in the type or race (phylogeny). Ontogeny was therefore a recapitulation of phylogeny, illustrating the so-called "biogenetic law." Change, therefore, was primarily *individual* variation that took place according to a preordained plan and pattern. This concept fitted in quite nicely with Agassiz's creationism. Contrary to some late nineteenth-century advocates of the evolution concept like Ernst Haeckel (1834–1919), it is not at all necessary for the validation of the idea of evolution.]

The classes of Crustacea and of Insects[110a] are particularly instructive in this respect. While the embryo of the highest Crustacea, the Brachyura, resembles by its form and structure the lowest types of this class, as the Entomostraca and Isopoda, it next assumes the shape of those of a higher order, the Macroura, before it appears with all the characteristics of the Brachyura.

Embryology furnishes also the best measure of the true affinities existing between animals. I do not mean to say that the affinities of animals can only be ascertained by embryonic investigations; the history of Zoology shows, on the contrary, that even before the study of the formation and growth of animals had become a distinct branch of physiology, the general relationship of most animals had already been determined with a remarkable degree of accuracy by anatomical investigations. It is nevertheless true that in some remarkable instances the knowledge of the embryonic changes of certain animals gave the first clue to their true affinities, while in other cases it has furnished a very welcome confirmation of relationships, which before could appear probable but were still very problematical. Even Cuvier considered, for instance, the Barnacles as a distinct class, which he placed among Mollusks, under the name of Cirripeds. It was not until Thompson had shown, what was soon confirmed by Burmeister and Martin St. Ange,[111] that the young Barnacle has a structure and form identical with that of some of the most common Entomostraca, that their true position in the system of animals could be determined; when they had to be removed to the class of Crustacea, among Articulata. The same was the case with the Lernæans, which Cuvier arranged with the Intestinal Worms, and which Nordmann has shown upon embryological evidence to belong also to the class of Crustacea.[112] Lamarck associated the Crinoids with Polypi, and though they were removed to the class of Echinoderms by Cuvier before the metamorphoses of the Comatula were known, the discovery of their pedunculated young furnished a direct proof that this was their true position.

[110a] It is expected that Embryology will furnish the means of ascertaining the relative standing of every family.

[111] John W. Thompson, *Zoological Researches* (6 pts., London, 1828–1834); Hermann Burmeister, *Beiträge zur Naturgeschichte der Rankenfüsser. Cirrepedia* (Berlin, 1834); G. J. Martin St. Ange, *Memoire sur l'organisation des Cirripèdes* (Paris, 1835).

[112] Alexander von Nordmann, *Micrographische Beiträge . . .* (2 vols., Berlin, 1832).

Embryology affords further a test for homologies in contradistinction of analogies. It shows that true homologies are limited respectively within the natural boundaries of the great branches of the animal kingdom.

The distinction between homologies and analogies, upon which the English naturalists have first insisted,[113] has removed much doubt respecting the real affinities of animals which could hardly have been so distinctly appreciated before. It has taught us to distinguish between real affinity based upon structural conformity, and similarity based upon mere external resemblance in form and habits. But even after this distinction had been fairly established, it remained to determine within what limits homologies may be traced. The works of Oken, Spix, Geoffroy and Carus,[114] show to what extravagant comparisons a preconceived idea of unity may lead. It was not until von Baer had shown that the development of the four great branches of the animal kingdom is essentially different,[115] that it could even be suspected that organs performing identical functions may be different in their essential relations to one another, and not until Rathke[116] had demonstrated that the yolk is in open communication with the main cavity of the Articulata on the dorsal side of the animal, and not on the ventral side, as in Vertebrata, that a solid basis was obtained for the natural limitation of true homologies. It now appears more and more distinctly, with every step of the progress Embryology is making, that the structure of animals is only homologous within

[113] Swainson, *Geography and Classification of Animals.*

[114] With reference to this point, consult: Lorenz Oken, *Ueber die Bedeutung der Schädel-Knochen* (Frankfurt, 1807), pamphlet, and *Lehrbuch der Naturphilosophie* (3 vols., Jena, 1809–1811; tr., A. Tulk, London, 1847); Spix, *Cephalogenesis, sive capitis ossei structura, formatio et significatio* (Munich, 1815); Étienne Geoffroy St. Hilaire, *Philosophie anatomique* (2 vols., Paris, 1818–1823); Carus, *Von den Ur-Theilen des Knochen-und Schalengerüstes* (Leipzig, 1828); Owen, *On the Archetype and Homologies of the Vertebrate Skeleton* (London, 1848); Cuvier, "Sur un nouveau rapprochement à établir entre les Classes qui composent le Règne animal," *Annales du Muséum d'Histoire Naturelle,* XIX (1812), 73; von Baer, *Entwickelungsgeschichte;* Leuckart, *Ueber die Morphologie und die Verwandtschaftsverhältnisse der wirbellosen Thiere* (Brunswick, 1848); Agassiz, *Twelve Lectures on Embryology.*

[115] Von Baer, *Entwickelungsgeschichte,* I, 160, 224. The extent of von Baer's information and the comprehensiveness of his views nowhere appear so strikingly as in this part of his work.

[116] Martin Heinrich Rathke, *Untersuchungen über die Bildung und Entwickelung des Flusskrebses* (Leipzig, 1829), and *Beiträge zur vergleichenden Anatomie und Physiologie, Reisebemerkungen aus Skandinavien* (Danzig, 1842).

the limits of the four great branches of the animal kingdom and that general homology, strictly proved, proves also typical identity, as special homology proves class identity.

The results of all embryonic investigations of modern times go to show more and more extensively that animals are entirely independent of external causes in their development. The identity of the metamorphoses of oviparous and viviparous animals belonging to the same type, furnishes the most convincing evidence to that effect.[117] Formerly it was supposed that the embryo could be affected directly by external influences to such an extent, that monstrosities, for instance, were ascribed to the influence of external causes. Direct observation has shown that they are founded upon peculiarities of the normal course of their development. The snug berth in which the young undergo their first transformation in the womb of their mother in all Mammalia excludes so completely the immediate influence of any external agent, that it is only necessary to allude to it to show how independent their growth must be of the circumstances in which even the mother may be placed. This is equally true of all other viviparous animals, as certain snakes, certain sharks, and the viviparous fishes. Again, the uniformity of temperature in the nests of birds and the exclusion, to a certain degree, of influences which might otherwise reach them in the various structures animals build for the protection of their young or of their eggs, show distinctly that the instinct of all animals leads them to remove their progeny from the influence of physical agencies, or to make these agents sub-

[117] This seems the most appropriate place to remark that the distinction made between viviparous and oviparous animals is not only untenable as far as their first origin in the egg is concerned, but also unphysiological, if it is intended by this designation to convey the idea of any affinity or resemblance in their respective modes of development. Fishes show more distinctly than any other class that animals, the development of which is identical in all leading features, may either be viviparous or oviparous; the difference here arising only from the connection in which the egg is developed and not from the development itself. Again, viviparous and oviparous animals of different classes differ greatly in their development, even though they may agree in laying eggs or bringing forth living young. The essential feature upon which any important generalization may be based is of course the mode of development of the germ. In this respect we find that Selachians, whether oviparous or viviparous, agree with one another; this is also the case with the bony fishes and the reptiles, whether they are respectively oviparous or viviparous; even the placentalian and non-placentalian Mammalia agree with one another in what is essential in their development. Too much importance has thus far been attached to the connections in which the germ is developed, to the exclusion of the leading features of the transformations of the germ itself.

servient to their purposes, as in the case of the ostrich. Reptiles and terrestrial Mollusks bury their eggs to subtract them from varying influences; fishes deposit them in localities where they are exposed to the least changes. Insects secure theirs in various ways. Most marine animals living in extreme climates lay their eggs in winter, when the variations of external influences are reduced to a minimum. Everywhere we find evidence that the phenomena of life, though manifested in the midst of all the most diversified physical influences, are rendered independent of them to the utmost degree by a variety of contrivances prepared by the animals themselves in self-protection, or for the protection of their progeny from any influence of physical agents not desired by them or not subservient to their own ends.

SECTION XIX

DURATION OF LIFE

There is the most extraordinary inequality in the average duration of the life of different kinds of animals and plants. While some grow and reproduce themselves and die in a short summer, nay, in a day, others seem to defy the influence of time.

Who has thus apportioned the life of all organized beings? To answer this question, let us first look at the facts of the case. In the first place, there is no conformity between the duration of life and either the size, or structure, or habitat of animals; next, the system in which the changes occurring during any period are regulated differs in almost every species, there being only a slight degree of uniformity between the representatives of different classes, within certain limits.

In most Fishes and the Reptiles proper, for instance, the growth is very gradual and uniform, and their development continues through life, so much so that their size is continually increasing with age.

In others, the Birds, for instance, the growth is rapid during the first period of their life until they have acquired their full size, and then follows a period of equilibrium, which lasts for a longer or shorter period in different species.

In others again, which also acquire within certain limits a definite size, the Mammalia, for instance, the growth is slower in early life, and maturity is attained as in man at an age which forms a much longer part of the whole duration of life.

In Insects the period of maturity is, on the contrary, generally the shortest, while the growth of the larva may be very slow, or at least that stage of development lasts for a much longer time than the life of the perfect Insects. There is no more striking example of this peculiar mode of growth than the seventeen-years' locust, so fully traced by Miss M. H. Morris.[118]

While all longlived animals continue as a matter of course their existence through a series of years under the varying influence of successive seasons, there are many others which are periodical in their appearance; this is the case with most insects, but perhaps in a still more striking manner with Medusæ.

The most interesting point in this subject is yet the change of character which takes place in the different stages of growth of one and the same animal. Neither Vertebrata, nor Mollusks, nor even Radiata exhibit in this respect anything so remarkable in the continuous changes which an individual animal may undergo as the Insects, and among them those with so-called complete metamorphoses, in which the young (the larva) may be an active, wormlike, voracious, even carnivorous being, which in middle life (the chrysalis) becomes a mummylike, almost motionless maggot, incapable of taking food, ending life as a winged and active insect. Some of these larvæ may be aquatic and very voracious, when the perfect insect is aërial and takes no food at all.

Is there any thing in this regulation of the duration of life in animals which recalls the agency of physical forces? Does not, on the contrary, the fact that while some animals are periodical and bound to the seasons in their appearance, others are independent of the course of the year, show distinctly their independence of all those influences which under a common expression are called physical causes? Is this not further illustrated in the most startling manner by the extraordinary changes, above alluded to, which one and the same animal may undergo during different periods of its life? Does this not prove directly the immediate intervention of a power

[118] See Harris, *Insects Injurious to Vegetation,* p. 184.

capable of controlling all these external influences, as well as regulating the course of life of every being, and establishing it upon such an immutable foundation within its cycle of changes, that the uninterrupted action of these agents does not interfere with the regular order of their natural existence?

There is, however, still another conclusion to be drawn from these facts: they point distinctly at a discriminating knowledge of time and space, at an appreciation of the relative value of unequal amounts of time and an unequal repartition of small, unequal periods over longer periods, which can only be the attribute of a thinking being.

SECTION XX

ALTERNATE GENERATIONS

While some animals go on developing gradually from the first formation of their germ to the natural end of their life and bring forth, generation after generation, a progeny which runs with never varying regularity through the same course, there are others which multiply in various ways, by division and by budding, or by a strange succession of generations, differing one from the other, and not returning by a direct course to their typical cycle.

The facts which have led to the knowledge of the phenomena now generally known under the name of *alternate generation* were first observed by Chamisso and Sars, and afterwards presented in a methodical connection by Steenstrup, in his famous pamphlet on that subject.[119] As a brief account of the facts may be found in almost every text-book of Physiology, I need not repeat them here, but only refer to the original investigations in which all the details known upon this subject may be found. These facts show, in the first place with regard to Hydroid Medusæ, that the individuals born from eggs, may be entirely different from those which produced the eggs, and end their life without ever undergoing themselves such changes as

[119] Ludovici A. de Chamisso, *De animalibus quibusdam e classe Vermium Linneana* (Berlin, 1819); Michael Sars, *Beskrivelser og Jagttagelser over nogle maerkelige eller nye i Havet ved den Bergenske Kyst levende Dyr* . . . (Bergen, 1835); Johann Steenstrup, *Ueber den Generations-Wechsel oder die Fortpflanzung und Entwickelung durch abwechselnde Generationen* . . . (Copenhagen, 1842).

would transform them into individuals similar to their parents;[120] and they show further, that this brood originating from eggs may increase and multiply by producing new individuals like themselves (*Syncoryne*), or of two kinds (*Campanularia*), or even individuals of various kinds, differing all to a remarkable extent, one from the other (*Hydractinia*), but in neither case resembling their common parent. None of these new individuals have distinct reproductive organs, any more than the first individuals born from eggs, their multiplication taking place chiefly by the process of budding; but as these buds remain generally connected with the first individual born from an egg, they form compound communities, similar to some polypstocks. Now some of these buds produce at certain seasons new buds of an entirely different kind, which generally drop off from the parent stock at an early period of their development (as in *Syncoryne, Campanularia,* etc.), and then undergo a succession of changes which end by their assuming the character of the previous egg-laying individuals, organs of reproduction of the two sexes developing meanwhile in them, which, when mature, lead to the production of new eggs; in others (as in *Hydractinia*) the buds of this kind do not drop off, but fade away upon the parent stock after having undergone all their transformations and also produced, in due time, a number of eggs.[121]

In the case of the Medusæ proper the parent lays eggs, from which originate polyplike individuals; but here these individuals divide by transverse constrictions into a number of disks, every one of which undergoes a succession of changes which end in the production of as many individuals, each identical with the parent, and capable in its turn of laying eggs (some, however, being males and others females). But the polyplike individuals born from eggs may also multiply by budding and each bud undergo the same changes as the first, the base of which does not die, but is also capable of growing up again and of repeating the same process.

[120] Polymorphism among individuals of the same species is not limited to Acalephs; it is also observed among genuine Polyps, the Madrepores, for example, and among Bryozoa, Ascidians, Worms, Crustacea (Lupea), and even among Insects (Bees).

[121] I have observed many other combinations of a similar character among the Hydroid Medusæ, which I shall describe at full length in my second volume and to which I do not allude here, as they could not be understood without numerous drawings. The case of Hydractinia is not quite correctly represented in the works in which that animal has been described. [See Agassiz, *Contributions,* IV (1862), 181–372.]

In other classes other phenomena of a similar character have been observed, which bear a similar explanation. J. Müller has most fully illustrated the alternate generations of the Echinoderms; Chamisso, Steenstrup, Eschricht, Krohn, and Sars, those of the Salpæ; von Siebold, Steenstrup, and others, those of certain Intestinal Worms.

This alternate generation differs essentially from metamorphosis, though some writers have attempted to identify these two processes. In metamorphosis as observed among Insects the individual born from an egg goes on undergoing change after change, in direct and immediate succession, until it has reached its final transformation; but however different it may be at different periods of its life, it is always one and the same individual. In alternate generations the individual born from an egg never assumes through a succession of transformations the characters of its parent, but produces, either by internal or external budding or by division, a number, sometimes even a large number of new individuals, and it is this progeny of the individuals born from eggs which grows to assume again the characters of the egg-laying individuals.

There is really an essential difference between the sexual reproduction of most animals and the multiplication of individuals in other ways. In ordinary sexual reproduction every new individual arises from an egg and by a regular succession of changes assumes the character of its parents. Now, though all species of animals reproduce their kind by eggs, and though in each there is at least a certain number of individuals, if not all, which have sprung from eggs, this mode of reproduction is not the only one observed among animals. We have already seen how new individuals may originate from buds, which in their turn may produce sexual individuals; we have also seen how, by division, individuals may also produce other individuals differing from themselves quite as much as the sexual buds, alluded to above, may differ from the individuals which produce them. There are yet still other combinations in the animal kingdom. In Polyps, for instance, every bud, whether it is freed from the parent stock or not, grows at once up to be a new sexual individual; while in many animals which multiply by division every new individual thus produced assumes at once the characters of those born from eggs. There is, finally, one mode of reproduction which is peculiar to certain Insects, in which several generations of fertile females follow one another before males appear again.

What comprehensive views must physical agents be capable of taking, and what a power of combination must they possess, to be able to ingraft all these complicated modes of reproduction upon structures already so complicated! — But if we turn away from mere fancies and consider the wonderful phenomena just alluded to in all their bearings, how instructive they appear with reference to this very question of the influence of physical agents upon organized beings! For here we have animals endowed with the power of multiplying in the most extraordinary ways, every species producing new individuals of its own kind, differing to the utmost from their parents. Does this not seem at first as if we had before us a perfect exemplification of the manner in which different species of animals may originate, one from the other, and increase the number of types existing at first? And yet, with all this apparent freedom of transformation, what do the facts finally show? That all these transformations are the successive terms of a cycle, as definitely closed within precise limits as in the case of animals, the progeny of which resembles for ever the immediate parent in all successive generations. For here, as everywhere in the organic kingdoms, these variations are only the successive expressions of a well regulated cycle, ever returning to its own type.

SECTION XXI

SUCCESSION OF ANIMALS AND PLANTS IN GEOLOGICAL TIMES

Geologists hardly seem to appreciate fully the whole extent of the intricate relations exhibited by the animals and plants whose remains are found in the different successive geological formations. I do not mean to say that the investigations we possess respecting the zoological and botanical characters of these remains are not remarkable for the accuracy and for the ingenuity with which they have been traced. On the contrary, having myself thus far devoted the better part of my life to the investigation of fossil remains, I have learned early, from the difficulties inherent in the subject, better to appreciate the wonderful skill, the high intellectual powers, the vast erudition displayed in the investigations of Cuvier and his successors upon the faunæ and floræ of past ages.[122] But I cannot refrain from expressing my wonder at the puerility of the discussions in which some geologists allow themselves still to indulge, in the face of such a vast amount of well digested facts as our science now possesses. They have hardly yet learned to see that there exists a definite order in the succession of these innumerable extinct beings; and of the relations of this gradation to the other great features exhibited by the animal kingdom, of the great fact that the development of life is the prominent trait in the history of our globe, they seem either to know nothing, or to look upon it only as a vague speculation, plausible perhaps, but hardly deserving the notice of sober science.

It is true, Palæontology as a science is very young; it has had to fight its course through the unrelenting opposition of ignorance and prejudice. What amount of labor and patience it has cost only to establish the fact that fossils are really the remains of animals and plants that once actually lived upon earth, only those know who are familiar with the history of science. Then it had to be proved that they are not the wrecks of the Mosaic deluge, which, for a time, was the prevailing opinion, even among scientific men.[123] After Cuvier had shown, beyond question, that they are the remains of animals no longer to be found upon earth among the living, Palæontology acquired for the first time a solid basis. Yet what an amount of labor it has cost to ascertain, by direct evidence, how these remains are distributed in the solid crust of our globe, what are the differences

[122] Cuvier, *Recherches sur les Ossemens fossiles . . .* ; James Sowerby, *The Mineral Conchology of Great Britain* (6 vols., London, 1812–1819); E. F. von Schlottheim, *Die Petrafactenkunde . . .* (Gotha, 1820), Lamarck, *Mémoires sur les fossiles des environs de Paris* (Paris, 1823); Georg A. Goldfuss, *Petrafacta Germaniæ* (9 pts., Düsseldorf, 1826–1833); Kaspar M. von Sternberg, *Versuch einer geognostisch-botanischen Darstellung der Flora der Vorwelt* (8 pts., Leipzig and Prague, 1820–1838); Alexandre Brongniart, *Prodrome d'une Histoire des Végétaux fossiles* (2 vols., Paris, 1818), and *Histoire des Végétaux fossiles* (2 vols., Paris, 1828–1843); John Lindley and William Hutton, *The Fossil Flora of Great Britain* (3 vols., London, 1831–1837); H. R. Göppert, *Systema Filicum fossilium* (Breslau and Bonn, 1836), *Die Gattungen der fossilen Pflanzen verglichen mit denen der Jetzwelt* (6 pts., Bonn, 1841–1848), and *Monographie der fossilen Coniferen* (Düsseldorf, 1850).

[123] Johann J. Scheuchzer, *Homo Diluvii testis . . .* (Zurich, 1726); Buckland, *Riliquiæ diluvianæ, or Observations on the Organic Remains . . . attesting the Action of an Universal Deluge* (London, 1824); August Scilla, *La vana speculazione desingannata del senso* (Naples, 1670).

they exhibit in successive formations,[124] what is their geographical distribution, only those can fully appreciate, who have had a hand in the work. And even now, how many important questions still await an answer!

One result, however, now stands unquestioned: the existence during each great geological era of an assemblage of animals and plants differing essentially for each period. And by period I mean those minor subdivisions in the successive sets of beds of rocks which constitute the stratified crust of our globe, the number of which is daily increasing, as our investigations become more extensive and more precise.[125] What remains to be done is to ascertain with more and more precision the true affinities of these remains to the animals and plants now living, the relations of those of the same period to one another and to those of the preceding and following epochs, the precise limits of these great eras in the development of life, the character of the successive changes the animal kingdom has undergone, the special order of succession of the representatives of each class, their combinations into distinct faunæ during each period, not to speak of the causes or even the circumstances under which these changes may have taken place.

In order to be able to compare the order of succession of the animals of past ages with some other prominent traits of the animal kingdom, it is necessary for me to make a few more remarks upon this topic. I can, fortunately, be very brief, as we possess a text-book of Palæontology arranged in zoological order, in which every one may at a glance see how, throughout all the classes of the animal kingdom, the different representatives of each in past ages are distributed in the successive geological formations.[126] From such a cursory survey it must appear that while certain types prevail during some periods, they are entirely foreign to others. This limitation is conspicuous with reference to entire classes among Vertebrata, while in other types it relates more to the orders or to the families and ex-

[124] See note 33, above.

[125] At first only three great periods were distinguished, the primary, the secondary, and the tertiary; afterwards, six or seven (Henry Thomas de la Bèche); later, from ten to twelve; now, the number is almost indefinite, at least undetermined in the present stage of our knowledge, when many geologists would only consider as subdivisions of longer periods what some paleontologists are inclined to consider as distinct periods.

[126] I allude to the classical work of François J. Pictet, *Traité élémentaire de Paléontologie* (4 vols., Paris, 1844–1845).

tends frequently only to the genera or the species. But whatever be the extent of their range in time, we shall see presently that all these types bear, as far as the order of their succession is concerned, the closest relation to the relative rank of living animals of the same types compared with one another, to the phrases of the embryonic growth of these types in the present day, and even to their geographical distribution upon the present surface of our globe. I will, however, select a few examples for further discussion. Among Echinoderms the Crinoids are for a long succession of periods the only representatives of that class; next follow the Starfish, and next the Sea-Urchins, the oldest of which belong to the type of Cidaris and Echinus, followed by Clypeastroids and Spatangoids. No satisfactory evidence of the existence of Holothuriæ has yet been found. Among Crustacea, while Trilobites are the only Crustacea of the oldest palæozoic rocks, there is found in the jurassic period a carcinological fauna entirely composed of Macrura, to which Brachyura are added in the tertiary period. The formations intermediate between the older palæozoic rocks and the Jura contain the remains of other Entomostraca, and later of some Macroura also. In both classes the succession of their representatives in different periods agrees with their respective standing, as determined by the gradation of their structure.

Among plants we find in the Carboniferous period prominently Ferns and Lycopodiaceæ; in the Triassic period Equisetaceæ and Coniferæ prevail; in the Jurassic deposits, Cycadeæ, and Monocotyledoneæ; while later only Dicotyledoneæ take the lead. The iconographic illustration of the vegetation of past ages has of late advanced beyond the attempts to represent the characteristic features of the animal world in different geological periods.

Without attempting here to characterize this order of succession, this much follows already from the facts mentioned, that while the material world is ever the same through all ages in all its combinations, as far back as direct investigations can trace its existence, organized beings, on the contrary, transform these same materials into ever new forms and new combinations. The carbonate of lime of all ages is the same carbonate of lime in form as well as composition, as long as it is under the action of physical agents only. Let life be introduced upon earth, and a Polyp builds its coral out of it, and each family, each genus, each species a different one, and different ones for all successive geological epochs. Phosphate of lime in palæozoic rocks is the same phosphate, as when prepared artificially by Man; but a Fish makes its spines out of it, and every Fish in its own way; Turtles their shield, Birds their wings, Quadrupeds their legs, and Man, like all other Vertebrates, his whole skeleton; and during each successive period in the history of our globe these structures are different for different species. What similarity is there between these facts? Do they not plainly indicate the working of different agencies excluding one another? Truly the noble frame of Man does not owe its origin to the same forces which combine to give a definite shape to the crystal. And what is true of the carbonate of lime is equally true of all inorganic substances; they present the same characters in all ages past as those they exhibit now.

Let us look upon the subject in still another light, and we shall see that the same is also true of the influence of all physical causes. Among these agents the most powerful is certainly electricity; the only one to which, though erroneously, the formation of animals has ever been directly ascribed. The effects it may now produce it has always produced, and produced them in the same manner. It has reduced metallic ores and various earthly minerals and deposited them in crystalline form, in veins, during all geological ages; it has transported these and other substances from one point to another in time past, as we may do now in our laboratories under its influence. Evaporation upon the surface of the earth has always produced clouds in the atmosphere, which after accumulating have been condensed in rain showers in past ages as now. Rain drop marks in the carboniferous and triassic rocks have brought to us this testimony of the identity of the operation of physical agents in past ages, to remind us that what these agents may do now they already did in the same way, in the oldest geological times, and have done at all times. Who, in the presence of such facts, could assume any causal connection between two series of phenomena, the one of which is ever obeying the same laws, while the other presents at every successive period new relations, an ever changing gradation of new combinations, leading to a final climax with the appearance of Man? Who does not see, on the contrary, that this identity of the products of physical agents in all ages totally disproves any influence on their part in the production of these ever changing beings which constitute the organic world, and which exhibit, as a whole, such striking evidence of connected thoughts!

SECTION XXII

LOCALIZATION OF TYPES IN PAST AGES

The study of the geographical distribution of the animals now living upon earth has taught us that every species of animals and plants has a fixed home, and even that peculiar types may be circumscribed within definite limits upon the surface of our globe. But it is only recently, since geological investigations have been carried on in remote parts of the world, that it has been ascertained that this special localization of types extends to past ages. Lund [127] for the first time showed that the extinct Fauna of the Brazils, during the latest period of a past age, consists of different representatives of the very same types now prevalent in that continent; Owen[128] has observed similar relations between the extinct Fauna of Australia and the types now living upon that continent.

If there is any naturalist left who believes that the Fauna of one continent may be derived from another portion of the globe, the study of these facts in all their bearing may undeceive him.

It is well known how characteristic the Edentata are for the present Fauna of the Brazils, for there is the home of the Sloths (*Bradypus*), the Tatous (*Dasypus*), the Ant-eaters (*Myrmecophaga*); there also have been found those extraordinary extinct genera, the Megatherium, the Mylodon, the Megalonyx, the Glyptodon, and the many other genera described by Dr. Lund and Professor Owen, all of which belong to this same order of Edentata. Some of these extinct genera of Edentata had also representatives in North America during the same geological period,[129] thus showing that though limited

[127] Peter V. Lund, *Blik paa Brasiliens Dyrverden för sidste Jordomvæltning* (Copenhagen, 1841).

[128] "On the Geographical Distribution of Extinct Mammalia," *Annals and Magazine of Natural History,* XVII (1846), 197.

[129] Joseph Leidy, *A Memoir on the Extinct Sloth Tribe of North America* (Smithsonian Contributions to Knowledge, VII, Washington, 1855).

within similar areas the range of this type has been different in different epochs.

Australia, at present almost exclusively the home of Marsupials, has yielded also a considerable number of equally remarkable species and two extinct genera of that type, all described by Owen in a report to the British Association, in 1844, and in Mitchell's *Expeditions into the Interior of Australia.*[130]

How far similar facts are likely to occur in other classes remains to be ascertained. Our knowledge of the geographical distribution of the fossil remains is yet too fragmentary to furnish any further data upon this point. It is, however, worthy of remark that though the types of the oldest geological periods had a much wider distribution than most recent families exhibit now, some families of fishes largely represented in the Devonian system of the Old World have not yet been noticed among the fossils of that period in America, as, for instance, the Cephalaspids, the Dipteri, and the Acanthodi. Again, of the many gigantic Reptiles of the Triassic and Oolitic periods, none are known to occur elsewhere except in Europe, and it can hardly be simply owing to the less extensive distribution of these formations in other parts of the world, since other fossils of the same formations are known from other continents. It is more likely that some of them at least are peculiar to limited areas of the surface of the globe, as even in Europe their distribution is not extensive.

Without, however, entering upon debatable ground, it remains evident that before the establishment of the present state of things peculiar types of animals, which were formerly circumscribed within definite limits, have continued to occupy the same or similar grounds in the present period, even though no genetic connection can be assumed between them, their representatives in these different formations not even belonging to the same genera. Such facts are in the most direct contradiction with any assumption that physical agents could have anything to do with their origin; for though their occurrence within similar geographical areas might at first seem to favor such a view, it must be borne in mind that these so localized beings are associated with other types which have a much wider range, and, what is still more significant, they belong to different geological peri-

[130] [T. L. Mitchell, *Three Expeditions into the Interior of Eastern Australia* . . . 2 vols., London, 1838.]

ods, between which great physical changes have undoubtedly taken place. Thus the facts indicate precisely the reverse of what the theory assumes: they prove a continued similarity of organized beings during successive geological periods, notwithstanding the extensive changes in the prevailing physical conditions which the country they inhabited may have undergone at different periods. In whatever direction this theory of the origin of animals and plants under the influence of physical agents is approached, it can nowhere stand a critical examination. Only the deliberate intervention of an Intellect, acting consecutively according to one plan can account for phenomena of this kind.

SECTION XXIII

LIMITATION OF SPECIES TO PARTICULAR GEOLOGICAL PERIODS

Without entering into a discussion respecting the precise limits within which this fact is true, there can no longer be any doubt that not only species but all other groups of animals and plants have a definite range of duration, as well as individuals.[131] The limits of this duration, as far as species are concerned, generally coincide with great changes in the physical conditions of the earth's surface;[132] though, strange to say, most of those investigators who would ascribe the origin of organized beings to the influence of such causes maintain also that species may extend from one period to another, which implies that these are not affected by such changes.[133]

When considering in general the limitation of species to particular geological periods, we might very properly disregard the question of the simultaneity of the successive appearance and disappearance of Faunæ as in no way affecting the result of the investigation, as long as it is universally conceded that there is no species known among the fossils which extends through an indefinite series of geo-

[131] Compare Sect. XIX.

[132] Élie de Beaumont, *Recherches sur quelquesunes des révolutions de la surface du globe* (Paris, 1830).

[133] For indications respecting the occurrence of all species of fossil organized beings now known, consult Heinrich G. Bronn, *Index palæontologicus* (3 vols., Stuttgart, 1848–1849); Alcide d'Orbigny, *Prodrome de Paléontologie stratigraphique universelle* (2 vols., Paris, 1850); John Morris, *Catalogue of the British Fossils* (London, 1854).

logical formations. Moreover, the number of the species, still considered as identical in several successive periods, is growing smaller and smaller in proportion as they are more closely compared. I have already shown, long ago, how widely many of the tertiary species, long considered as identical with living ones, differ from them, and also how different the species of the same family may be in successive subdivisions of the same great geological formation.[134] Hall has come to the same result in his investigations of the fossils of the State of New York.[135] Every monograph reduces their number, in every formation. Thus Barrande, who has devoted so many years to the most minute investigation of the Trilobites of Bohemia, has come to the conclusion that their species do not extend from one formation to the other; d'Orbigny and Pictet have come to the same conclusion for the fossil remains of all classes. It may well be said that as fossil remains are studied more carefully in a zoological point of view, the supposed identity of species in different geological formations vanishes gradually more and more; so that the limitation of species in time, already ascertained in a general way by the earlier investigations of their remains in successive geological formations, is circumscribed step by step within narrower, more definite, and also more equable periods. Species are truly limited in time, as they are limited in space, upon the surface of the globe. The facts do not exhibit a gradual disappearance of a limited number of species and an equally gradual introduction of an equally limited number of new ones; but on the contrary, the simultaneous creation and the simultaneous destruction of entire faunæ, and a coincidence between these changes in the organic world and the great physical changes our earth has undergone. Yet it would be premature to attempt to determine the extent of the geographical range of these changes, and still more questionable to assert their synchronism upon the whole surface of the globe in the ocean and upon dry land.

To form adequate ideas of the great physical changes the surface of our globe has undergone, and the frequency of these modifications of the character of the earth's surface, and of their coincidence with the changes observed among the organized beings, it is necessary to

[134] Agassiz, *Coquilles tertiares reputées identiques avec les espèces vivantes* (Neuchâtel, 1845); and *Études critiques sur les Mollusques fossiles* (4 vols., Neuchâtel, 1840–1845).

[135] *Palæontology of New York.* [See above, n. 33.]

study attentively the works of Élie de Beaumont.[136] He for the first time attempted to determine the relative age of the different systems of mountains, and showed first also that the physical disturbances occasioned by their upheaval coincided with the successive disappearance of entire faunæ and the reappearance of new ones. In his earlier papers he recognized seven, then twelve, afterwards fifteen such great convulsions of the globe, and now he has traced more or less fully and conclusively the evidence that the number of these disturbances has been at least sixty, perhaps one hundred. But while the genesis and genealogy of our mountain systems were thus illustrated, palæontologists, extending their comparisons between the fossils of different formations more carefully to all the successive beds of each great era, have observed more and more marked differences between them and satisfied themselves that faunæ also have been more frequently renovated than was formerly supposed; so that the general results of geology proper and of palæontology concur in the main to prove that, while the globe has been at repeated intervals and indeed frequently, though after immensely long periods, altered and altered again, until it has assumed its present condition, so have also animals and plants living upon its surface been again and again extinguished and replaced by others, until those now living were called into existence with man at their head. The investigation is not in every case sufficiently complete to show everywhere a coincidence between this renovation of animals and plants and the great physical revolutions which have altered the general aspect of the globe, but it is already extensive enough to exhibit a frequent synchronism and correlation, and to warrant the expectation that it will in the end lead to a complete demonstration of their mutual dependence, not as cause and effect, but as steps in the same progressive development of a plan which embraces the physical as well as the organic world.

In order not to misapprehend the facts and perhaps to fall back upon the idea that these changes may be the cause of the differences observed between the fossils of different periods, it must be well understood that while organized beings exhibit through all geological formations a regular order of succession, the character of which will be more fully illustrated hereafter, this succession has been from time to time violently interrupted by physical disturbances, without

[136] *Notice sur les systèmes de montagnes* (Paris, 1852).

any of these altering in any way the progressive character of that succession of organized beings. Truly this shows that the important, the leading feature of this whole drama is the development of life, and that the material world affords only the elements for its realization. The simultaneous disappearance of entire faunæ and the following simultaneous appearance of other faunæ show further that, as all these faunæ consist of the greatest variety of types in all formations, combined everywhere into natural associations of animals and plants between which there have been definite relations at all times, their origin can at no time be owing to the limited influence of monotonous physical causes ever acting in the same way. Here again, the intervention of a Creator is displayed in the most striking manner, in every stage of the history of the world.

SECTION XXIV

PARALLELISM BETWEEN THE GEOLOGICAL SUCCESSION OF ANIMALS AND PLANTS AND THEIR PRESENT RELATIVE STANDING

The total absence of the highest representatives of the animal kingdom in the oldest deposits forming part of the crust of our globe has naturally led to the very general belief that the animals which have existed during the earliest period of the history of our earth were inferior to those now living, nay, that there is a natural gradation from the oldest and lowest animals to the highest now in existence. To some extent this is true; but it is certainly not true that all animals form one simple series from the earliest times, during which only the lowest types of animals would have been represented, to the last period, when Man appeared at the head of the animal creation.[137] It has already been shown (Sect. VII.) that representatives of all the great types of the animal kingdom have existed from the beginning of the creation of organized beings. It is therefore not in the successive appearance of the great branches of the animal kingdom that we may expect to trace a parallelism between their succession in geological times and their relative standing at present. Nor can any such correspondence be observed between the appearance of classes, at

[137] Agassiz, *Twelve Lectures on Embryology,* pp. 68, 128.

least not among Radiata, Mollusks, and Articulata, as their respective classes seem to have been introduced simultaneously upon our earth, with perhaps the sole exception of the Insects, which are not known to have existed before the Carboniferous period. Among Vertebrata, however, there appears already a certain coincidence, even within the limits of the classes, between the time of their introduction and the rank their representatives hold in comparison to one another. But upon this point more hereafter.

It is only within the limits of the different orders of each class that the parallelism between the succession of their representatives in past ages and their respective rank in the present period is decidedly characteristic. But if this is true, it must be at the same time obvious to what extent the recognition of this correspondence may be influenced by the state of our knowledge of the true affinities and natural gradation of living animals, and that until our classifications have become the correct expression of these natural relations even the most striking coincidence with the succession of their representatives in past ages may be entirely overlooked. On that account it would be presumptuous on my part to pretend that I could illustrate this proposition through the whole animal kingdom, as such an attempt would involve the assertion that I know all these relations, or that where there exists a discrepancy between the classification and the succession of animals the classification must be incorrect, or the relationship of the fossils incorrectly appreciated. I shall therefore limit myself here to a general comparison, which may, however, be sufficient to show that the improvements which have been introduced in our systems upon purely zoological grounds have nevertheless tended to render more apparent the coincidence between the relative standing among living animals and the order of succession of their representatives in past ages. I have lately attempted to show that the order of Halcyonoids among Polyps is superior to that of Actinoids; that in this class compound communities constitute a higher degree of development, when contrasted with the characters and mode of existence of single Polyps, as exhibited by the Actinia; that top-budding is superior to lateral budding; and that the type of Madrepores, with their top-animal, or at least with a definite and limited number of tentacles, is superior to all other Actinoids. If this be so, the prevalence of Actinoids in older geological formations, to the exclusion of

Halcyonoids, the early prevalence of Astræoids, and the very late introduction of Madrepores, would already exhibit a correspondence between the rank of the living Polyps and the representatives of that class in past ages, though we may hardly expect a very close coincidence in this respect between animals the structure of which is so simple. The prevalence of *Rugosa* and *Tabulata* in the oldest deposits appears in a new light, since it has been known that the *Tabulata* are Hydroids, and not genuine Polyps.

The gradation among the orders of Echinoderms is perfectly plain. Lowest stand the Crinoids, next the Asterioids, next the Echinoids, and highest the Holothurioids. Ever since this class has been circumscribed within its natural limits, this succession has been considered as expressing their natural relative standing, and modern investigations respecting their anatomy and embryology, however extensive, have not led to any important change in their classification, as far as the estimation of their rank is concerned. This is also precisely the order in which the representatives of this class have successively been introduced upon earth in past geological ages. Among the oldest formations we find pedunculated Crinoids only, and this order remains prominent for a long series of successive periods; next come free Crinoids and Asterioids; next Echinoids, the successive appearance of which since the Triassic period to the present day coincides also with the gradation of their subdivisions, as determined by their structure; and it was not until the present period that the highest Echinoderms, the Holothurioids, have assumed a prominent position in their class.

Among Acephala there is not any more uncertainty respecting the relative rank of their living representatives than among Echinoderms. Every zoologist acknowledges the inferiority of the Bryozoa and the Brachiopods when compared with the Lamellibranchiata, and among these the inferiority of the Monomyaria in comparison with the Dimyaria would hardly be denied. Now if any fact is well established in Palæontology it is the earlier appearance and prevalence of Bryozoa and Brachiopods in the oldest geological formations and their extraordinary development for a long succession of ages, until Lamellibranchiata assume the ascendancy which they maintain to the fullest extent at present. A closer comparison of the different families of these orders might further show how close this correspondence is through all ages.

Of Gasteropoda I have nothing special to say, as every palæontologist is aware how imperfectly their remains have been investigated in comparison with what has been done for the fossils of other classes. Yet the Pulmonata are known to be of more recent origin than the Branchifera, and among these the Siphonostomata to have appeared later than the Holostomata, and this exhibits already a general coincidence between their succession in time and their respective rank.

Our present knowledge of the anatomy of the Nautilus, for which science is indebted to the skill of Owen,[138] may satisfy everybody that among Cephalopods the Tetrabranchiata are inferior to the Dibranchiata; and it is not too much to say that one of the first points a collector of fossils may ascertain for himself is the exclusive prevalence of the representatives of the first of these types in the oldest formations, and the later appearance, about the middle geological ages, of representatives of the other type, which at present is the most widely distributed.

Of Worms nothing can be said of importance with reference to our inquiry; but the Crustacea exhibit, again, the most striking coincidence. Without entering into details, it appears from the classification of Milne-Edwards[139] that Decapods, Stomapods, Amphipods, and Isopods constitute the higher orders, while Branchiopods, Entomostraca, Trilobites, and the parasitic types, constitute, with Limulus, the lower orders of this class. In the classification of Dana his first type embraces Decapods and Stomapods, the second Amphipods and Isopods, the third Entomostraca, including Branchiopods, the fourth Cirripedia, and the fifth Rotatoria. Both acknowledge in the main the same gradation; though they differ greatly in the combination of the leading groups, and also the exclusion by Milne-Edwards of some types, as the Rotifera, which Burmeister first, then Dana and Leydig,[140] unite justly, as I believe, with the Crustacea. This gradation now presents the most perfect coincidence with the order of succession of Crustacea in past geological ages, even down to their sub-

[138] *Memoir on the Pearly Nautilus* (London, 1832).

[139] Henri Milne-Edwards, *Histoire naturelle des Crustacés* (3 vols., Paris, 1834–1840).

[140] Dana, *Crustacea,* p. 45; Franz Leydig, "Räderthiere . . . ," *Zeitschrift f. Wissenschaften Zoologie,* VI (1854), 1.

divisions into minor groups. Trilobites and Entomostraca are the only representatives of the class in palæozoic rocks; in the middle geological ages appear a variety of Shrimps, among which the Macrouran Decapods are prominent, and later only the Brachyoura, which are the most numerous in our days.

The fragmentary knowledge we possess of the fossil Insects does not justify us yet in expecting to ascertain with any degree of precision the character of their succession through all geological formations, though much valuable information has already been obtained respecting the entomological faunæ of several geological periods.

The order of succession of Vertebrata in past ages exhibits features in many respects differing greatly from the Articulata, Mollusks, and Radiata. Among these we find their respective classes appearing simultaneously in the oldest periods of the history of our earth. Not so with the Vertebrata, for though Fishes may be as old as any of the lower classes, Reptiles, Birds, and Mammalia are introduced successively in the order of their relative rank in their types. Again, the earliest representatives of these classes do not always seem to be the lowest; on the contrary, they are to a certain extent and in a certain sense the highest, in as far as they embody characters which in later periods appear separately in higher classes (see Sect. XXVI) to the exclusion of what henceforth constitutes the special character of the lower class. For instance, the oldest Fishes known partake of the characters which at a later time are exclusively found in Reptiles and no longer belong to the Fishes of the present day. It may be said that the earliest Fishes are rather the oldest representatives of the type of Vertebrata than of the class of Fishes, and that this class assumes only its proper characters after the introduction of the class of Reptiles upon earth. Similar relations may be traced between the Reptiles and the classes of Birds and Mammalia, which they precede. I need only allude here to the resemblance of the Pterodactyli and the Birds, and to that of Ichthyosauri and certain Cetacea. Yet through all these intricate relations there runs an evident tendency towards the production of higher and higher types, until at last Man crowns the whole series. Seen as it were at a distance, so that the mind can take a general survey of the whole and perceive the connection of the successive steps without being bewildered by the details, such a series appears like the development of a great conception, expressed in such harmonious proportions that every link appears necessary to the full comprehension of its meaning; and yet so independent and perfect in itself, that it might be mistaken for a complete whole; and again, so intimately connected with the preceding and following members of the series, that one might be viewed as flowing out of the other. What is universally acknowledged as characteristic of the highest conceptions of genius is here displayed in a fulness, a richness, a magnificence, an amplitude, a perfection of details, a complication of relations, which baffle our skill and our most persevering efforts to appreciate all its beauties. Who can look upon such series, coinciding to such an extent, and not read in them the successive manifestations of a thought, expressed at different times, in ever new forms, and yet tending to the same end, onwards to the coming of Man, whose advent is already prophesied in the first appearance of the earliest Fishes!

The relative standing of plants presents a somewhat different character from that of animals. Their great types are not built upon so strictly different plans of structure; they exhibit, therefore, a more uniform gradation from their lowest to their highest types, which are not personified in one highest plant, as the highest animals are in Man.

Again, Zoology is more advanced respecting the limitation of the most comprehensive general divisions than Botany, while Botany is in advance respecting the limitation and characteristics of families and genera. There is on that account more diversity of opinion among botanists respecting the number and the relative rank of the primary divisions of the vegetable kingdom than among zoologists respecting the great branches of the animal kingdom. While most writers agree in admitting among plants such primary groups as Acotyledones, Monocotyledones, and Dicotyledones under these or other names, others would separate the Gymnosperms from the Dicotyledones.

It appears to me that this point in the classification of the living plants cannot be fully understood without a thorough acquaintance with the fossils and their distribution in the successive geological formations, and that this case exhibits one of the most striking examples of the influence classification may have upon our appreciation of the gradation of organized beings in the course of time. As long as

Gymnosperms stand among Dicotyledones, no relation can be traced between the relative standing of living plants and the order of succession of their representatives in past ages. On the contrary, let the true affinity of Gymnosperms with Ferns, Equisetaceæ, and especially with Lycopodiaceæ be fully appreciated, and at once we see how the vegetable kingdom has been successively introduced upon earth in an order which coincides with the relative position its primary divisions bear to one another, in respect to their rank, as determined by the complication of their structure. Truly, the Gymnosperms, with their imperfect flower, their open carpels supporting their polyembryonic seeds in their axis, are more nearly allied to the anathic Acrophytes with their innumerable spores, than to either the Monocotyledones or Dicotyledones; and, if the vegetable kingdom constitutes a graduated series beginning with Cryptogams, followed by Gymnosperms, and ending with Monocotyledones and Dicotyledones, have we not in that series the most striking coincidence with the order of succession of Cryptogams in the oldest geological formations, especially with the Ferns, Equisetaceæ, and Lycopodiaceæ of the Carboniferous period, followed by the Gymnosperms of the Trias and Jura and the Monocotyledones of the same formation and the late development of Dicotyledones? Here, as everywhere, there is but one order, one plan in nature.

SECTION XXV

PARALLELISM BETWEEN THE GEOLOGICAL SUCCESSION OF ANIMALS AND THE EMBRYONIC GROWTH OF THEIR LIVING REPRESENTATIVES

Several authors have already alluded to the resemblance which exists between the young of some of the animals now living and the fossil representatives of the same families in earlier periods. But these comparisons have thus far been traced only in isolated cases and have not yet led to a conviction that the character of the succession of organized beings in past ages is such in general as to show a remarkable agreement with the embryonic growth of animals; though the state of our knowledge in Embryology and Palæontology justifies now such a conclusion. The facts most important to a proper appreciation of this point have already been considered in the preceding paragraph, as far as they relate to the order of succession of animals, when compared with the relative rank of their living representatives. In examining now the agreement between this succession and the phases of the embryonic growth of living animals, we may therefore take for granted that the order of succession of their fossil representatives is sufficiently present to the mind of the reader to afford a satisfactory basis of comparison. Too few Corals have been studied embryologically to afford extensive means of comparison; yet so much is known, that the young polyp when hatched is an independent, simple animal, that it is afterwards incased in a cup secreted by the foot of the actinoid embryo, which may be compared to the external wall of the *Rugosa,* and that the polyp gradually widens until it has reached its maximum diameter, prior to budding or dividing; while in ancient corals this stage of enlargement seems to last during their whole life, as, for example, in the Cyathophylloids.[140a] None of the ancient Corals form those large communities, composed of myriads of united individuals, so characteristic of our coral reefs; the more isolated and more independent character of the individual polyps of past ages presents a striking resemblance to the isolation of young corals in all the living types. In no class is there, however, so much to learn still as in Polypi, before the correspondence of their embryonic growth and their succession in time can be fully appreciated. In this connection I would also remark that among the lower animals it is rarely observed that anyone, even the highest type, represents in its metamorphoses all the stages of the lower types, neither in their development nor in the order of their succession; and that frequently the knowledge of the embryology of several types of different standing is required, to ascertain the connection of the whole series in both spheres.

[140a] Since I have ascertained that the Tabulata are Hydroids and not Polyps, I have had my doubts respecting the real affinities of the Rugosa. The tendency to a quadripartite arrangement of their septa indicates unquestionably a nearer relation to Acalephs than to Polyps. Moreover, their successive floors are different from the interseptal floors of the true Polyps, and resemble those of the Tabulata. It may be, therefore, that their true affinity is rather with the Acalephs than with the Polyps, and that the family of Lucernaria is a living representative of that type, but without hard parts. In this case the foot-secretion of the Actinoids would only indicate a typical resemblance between Polyps and Acalephs, and not constitute an evidence of the relative standing of the two types.

No class, as yet, affords a more complete and more beautiful evidence of the correspondence of their embryonic changes with the successive appearance of their representatives in past ages than the Echinoderms, thanks to the extensive and patient investigations of J. Müller upon the metamorphoses of these animals. Prior to the publication of his papers the metamorphosis of the European Comatula alone was known. (See Sect. XVIII.) This had already shown that the early stages of growth of this Echinoderm exemplify the pedunculated Crinoids of past ages. I have myself seen further that the successive stages of the embryonic growth of Comatula typify, as it were, the principal forms of Crinoids which characterize the successive geological formations; first, it recalls the Cistoids of the palæozoic rocks, which are represented in its simple sphæroidal head; next the few-plated Platycrinoids of the Carboniferous period; next the Pentacrinoids of the Lias and Oolite, with their whorls of cirrhi; and finally, when freed from its stem, it stands as the highest Crinoid, as the prominent type of the family in the present period. The investigations of Müller upon the larvæ of all the families of living Asterioids and Echinoids enable us to extend these comparisons to the higher Echinoderms also. The first point which strikes the observers in the facts ascertained by Müller is the extraordinary similarity of so many larvæ, of such different orders and different families as the Ophiuroids and Asterioids, the Echinoids proper and the Spatangoids, and even the Holothurioids, all of which end of course in reproducing their typical peculiarities. It is next very remarkable that the more advanced larval state of Echinoids and Spatangoids should continue to show such great similarity, that a young Amphidetus hardly differs from a young Echinus. Finally, not to extend these remarks too far, I would only add that these young Echinoids (Spatangus, as well as Echinus proper) have rather a general resemblance to Cidaris on account of their large spines, than to Echinus proper. Now these facts agree exactly with what is known of the successive appearance of Echinoids in past ages; their earliest representatives belong to the genera Diadema and Cidaris, next come true Echinoids, later only Spatangoids. When the embryology of the Clypeastroids is known, it will no doubt afford other links to connect a larger number of the members of this series.

What is known of the embryology of Acephala, Gasteropoda, and

Cephalopoda affords but a few data for such comparisons. It is, nevertheless, worthy of remark, that while the young *Lamellibranchiata* are still in their embryonic stage of growth, they resemble, externally at least, Brachiopods more than their own parents, and the young shells of all Gasteropods known in their embryonic stage of growth, being all holostomate, recall the oldest types of that class. Unfortunately, nothing is yet known of the embryology of the Chambered Cephalopoda, which are the only ones found in the older geological formations, and the changes which the shield of the Dibranchiata undergoes have not yet been observed, so that no comparisons can be established between them and the Belemnites and other representatives of this order in the middle and more recent geological ages.

Respecting Worms, our knowledge of the fossils is too fragmentary to lead to any conclusion, even should our information of the embryology of these animals be sufficient as a basis for similar comparisons. The class of Crustacea, on the contrary, is very instructive in this respect; but to trace our comparisons through the whole series it is necessary that we should consider simultaneously the embryonic growth of the higher Entomostraca, such as Limulus, and that of the highest order of the class, when it will appear that as the former recall in early life the form and character of the Trilobites, so does the young Crab, passing through the form of the Isopods and that of the Macrouran Decapods before it assumes its typical form as Brachyouran, recall the well-known succession of Crustacea through the geological middle ages and the tertiary periods to the present day. The early appearance of Scorpions in the Carboniferous period is probably also a fact to the point, if, as I have attempted to show, Arachnidians may be considered as exemplifying the chrysalis stage of development of Insects; but for reasons already stated (Sect. XXIV) it is hardly possible to take Insects into consideration in these inquiries.

In my researches upon fossil Fishes I have pointed out at length the embryonic character of the oldest fishes, but much remains to be done in that direction. The only fact of importance I have learned of late is that the young Lepidosteus, long after it has been hatched, exhibits in the form of its tail, characters thus far only known among the fossil fishes of the Devonian system. It is to be hoped, that the embryology of the Crocodile will throw some light upon the succession of the gigantic Reptiles of the middle geological ages, as I shall show that the embryology of Turtles throws light upon the fossil Chelonians. It is already plain that the embryonic changes of Batrachians coincide with what is known of their succession in past ages. The fossil Birds are too little known, and the fossil Mammalia do not extend through a sufficiently long series of geological formations to afford many striking points of comparison; yet the characteristic peculiarities of their extinct genera exhibit everywhere indications that their living representatives in early life resemble them more than they do their own parents. A minute comparison of a young elephant with any mastodon will show this most fully, not only in the peculiarities of their teeth, but even in the proportion of their limbs, their toes, etc.

It may therefore be considered as a general fact, very likely to be more fully illustrated as investigations cover a wider ground, that the phases of development of all living animals correspond to the order of succession of their extinct representatives in past geological times. As far as this goes, the oldest representatives of every class may then be considered as embryonic types of their respective orders or families among the living. Pedunculated Crinoids are embryonic types of the Comatuloids, the oldest Echinoids embryonic representatives of the higher living families, Trilobites embryonic types of Entomostraca, the Oolitic Decapods embryonic types of our Crabs, the Heterocercal Ganoids embryonic types of the Lepidosteus, the *Andrias Scheuchzeri* an embryonic prototype of our Batrachians, the Zeuglodonts embryonic Sirenidæ, the Mastodons embryonic Elephants, etc.

To appreciate, however, fully and correctly all these relations, it is further necessary to make a distinction between embryonic types in general, which represent in their whole organization early stages of growth of higher representatives of the same type, and *embryonic features* prevailing more or less extensively in the characters of allied genera, as in the case of the Mastodon and Elephant, and what I would call *hypembryonic types,* in which embryonic features are developed to extremes in the further periods of growth, as, for instance, the wings of the Bats, which exhibit the embryonic character of a webbed hand, as all Mammalia have it at first, but here grown out and developed into an organ of flight, or assuming in other families the shape of a fin, as in the Whale, or the Sea-turtle, in which the close connection of the fingers is carried out to another extreme.

Without entering into further details upon this subject, which will be fully illustrated in my *Contributions to the Natural History of the United States,* enough has already been said to show that the leading thought which runs through the succession of all organized beings in past ages is manifested again in new combinations, in the phases of the development of the living representatives of these different types. It exhibits everywhere the working of the same creative Mind, through all times, and upon the whole surface of the globe.

SECTION XXVI

PROPHETIC TYPES AMONG ANIMALS

We have seen in the preceding Section, how the embryonic conditions of higher representatives of certain types, called into existence at a later time, are typified, as it were, in representatives of the same types which have existed at an earlier period. These relations, now they are satisfactorily known, may also be considered as exemplifying, as it were, in the diversity of animals of an earlier period the pattern upon which the phases of the development of other animals of a later period were to be established. They appear now, like a prophecy in those earlier times, of an order of things not possible with the earlier combinations then prevailing in the animal kingdom, but exhibiting in a later period, in a striking manner, the antecedent considerations of every step in the gradation of animals.

This is, however, by no means the only nor even the most remarkable case of such prophetic connections between facts of different dates.

Recent investigations in Palæontology have led to the discovery of relations between animals of past ages and those now living which were not even suspected by the founders of that science. It has, for instance, been noticed that certain types which are frequently prominent among the representatives of past ages combine in their structure peculiarities which at later periods are only observed separately in different, distinct types. Sauroid Fishes before Reptiles, Pterodactyles before Birds, Ichthyosauri before Dolphins, etc.

There are entire families among the representatives of older periods of nearly every class of animals, which in the state of their perfect development exemplify such prophetic relations, and afford within the limits of the animal kingdom, at least, the most unexpected evidence that the plan of the whole creation had been maturely considered long before it was executed. Such types I have for some time past been in the habit of calling *prophetic types*. The Sauroid Fishes of the past geological ages are an example of this kind. These Fishes, which have preceded the appearance of Reptiles, present a combination of ichthyic and reptilian characters not to be found in the true members of this class, which form its bulk at present. The Pterodactyles which have preceded the class of Birds, and the Ichthyosauri which have preceded the appearance of the Crustacea are other examples of such prophetic types. These cases suffice for the present to show that there is a real difference between *embryonic* types and *prophetic* types. Embryonic types are in a measure also prophetic types, but they exemplify only the peculiarities of development of the higher representatives of their own types; while prophetic types exemplify structural combinations observed at a later period in two or several distinct types and are, moreover, not necessarily embryonic in their character, as for example, the Monkeys in comparison to Man; while they may be so, as in the case of the Pinnate, Plantigrade, and Digitigrade Carnivora, or still more so in the case of the pedunculated Crinoids.

Another combination is also frequently observed among animals when a series exhibits such a succession as exemplifies a natural gradation, without immediate or necessary reference to either embryonic development or succession in time, as the Chambered Cephalopods. Such types I call *progressive types*.[141]

Again: a distinction ought to be made between prophetic types proper and what I would call *synthetic types,* though both are more or less blended in nature. Prophetic types proper are those which in

[141] Agassiz, "Progressive, Embryonic, and Prophetic Types," *Proceedings,* AAAS, II (1850), 432–438.

their structural complications lean towards other combinations fully realized in a later period, while synthetic types are those which combine in a well balanced measure features of several types occurring as distinct, only at a later time. Sauroid Fishes and Ichthyosauri are more distinctly synthetic than prophetic types, while Pterodactyles have more the character of prophetic types; so are also Echinocrinus with reference to Echini, Pentremites with reference to Asterioids, and Pentacrinus with reference to Comatula. Full illustrations of these different cases will yet be needed to render obvious the importance of such comparisons, and I shall not fail to present ample details upon this subject in my *Contributions to the Natural History of the United States*. Enough, however, has already been said to show that the character of these relations among animals of past ages, compared with those of later periods or of the present day, exhibits more strikingly than any other feature of the animal kingdom the thoughtful connection which unites all living beings through all ages into one great system, intimately linked together from beginning to end.

SECTION XXVII

PARALLELISM BETWEEN THE STRUCTURAL GRADATION OF ANIMALS AND THEIR EMBRYONIC GROWTH

So striking is the resemblance of the young of higher animals to the full-grown individuals of lower types, that it has been assumed by many writers that all the higher animals pass, during the earlier stages of their growth, through phases corresponding to the permanent constitution of the lower classes. These suppositions, the results of incomplete investigations, have even become the foundation of a system of philosophy of Nature, which represents all animals as the different degrees of development of a few primitive types.[142] These views have been too generally circulated of late in an anonymous

[142] Lamarck, *Philosophie zoologique;* Maillet (Pseudon. Telliamed), *Entretiens d'un philosophe Indien avec un missionnaire français sur la diminution de la mer, la formation de la terre, l'origine de l'homme . . .* (2 vols., Amsterdam, 1748); Oken, *Lehrbuch der Naturphilosophie; The Vestiges of the Natural History of Creation* (London, 1844).

work entitled *Vestiges of Creation*[143] to require further mention here. It has also been shown above (Sect. VIII) that animals do not form such a simple series as would result from a successive development. There remains therefore only for us to show now within what limits the natural gradation which may be traced in the different types of the animal kingdom corresponds to the changes they undergo during their growth, having already considered the relations which exist between these metamorphoses and the successive appearance of animals upon earth, and between the latter and the structural gradation or relative standing of their living representatives. Our knowledge of the complication of structure of all animals is sufficiently advanced to enable us to select, almost at random, our examples of the correspondence between the structural gradation of animals and their embryonic growth in all those classes the embryologic development of which has been sufficiently investigated. Yet in order to show more distinctly how closely all the leading features of the animal kingdom are combined, whether we consider the complication of their structure, or their succession in time, or their embryonic development, I shall refer by preference to the same types which I have chosen before for the illustration of the other relations.

Among Echinoderms, we find in the order of Crinoids the pedunculated types standing lowest, Comatulæ highest, and it is well known that the young Comatula is a pedunculated Crinoid, which only becomes free in later life. J. Müller has shown that among the Echinoids, even the highest representatives, the Spatangoids, differ but slightly in early youth from the Echinoids, and no zoologist can doubt that these are inferior to the former. Among Crustacea, Dana has insisted particularly upon the serial gradation which may be traced between the different types of Decapods, their order being naturally from the highest Brachyura, through the Anomoura, the Macroura, the Tetradecapods, etc., to the Entomostraca; the Macrouran character of the embryo of our Crabs has been fully illustrated by Rathke, in his beautiful investigations upon the embry-

[143] [Robert Chambers was the author of *Vestiges of the Natural History of Creation* and *Explanations: a Sequel to "Vestiges of the Natural History of Creation"* (London, 1846), books whose theories of change and development were opposed by the great majority of professional naturalists including those who would later support Darwin. See the preface by Alexander Ireland to the 1884 (London) edition of the *Vestiges*.]

ology of Crustacea. I have further shown that the young of Macroura represents even Entomostraca forms, some of these young having been described as representatives of that order. The correspondence between the gradation of Insects and their embryonic growth I have illustrated fully in a special paper.[144] Similar comparisons have been made in the class of Fishes; among Reptiles, we find the most striking examples of this kind among Batrachians (see above, Sect. XII); among Birds[145] the uniformly webbed foot in all young exhibits another correspondence between the young of higher orders and the permanent character of the lower ones. In the order of Carnivora, the Seals, the Plantigrades, and the Digitigrades exemplify the same coincidence between higher and higher representatives of the same types, and the embryonic changes through which the highest pass successively.

No more complete evidence can be needed to show that there exists throughout the animal kingdom the closest correspondence between the gradation of their types and the embryonic changes their respective representatives exhibit throughout. And yet what genetic relation can there exist between the Pentacrinus of the West Indies and the Comatulæ, found in every sea; what between the embryos of Spatangoids and those of Echinoids, and between the former and the adult Echinus; what between the larva of a Crab and our Lobsters; what between the Caterpillar of a Papilio and an adult Tinea, or an adult Sphinx; what between the Tadpole of a Toad and our Menobranchus; what between a young Dog and our Seals, unless it be the plan designed by an intelligent Creator?

SECTION XXVIII

RELATIONS BETWEEN THE STRUCTURE, THE EMBRYONIC GROWTH, THE GEOLOGICAL SUCCESSION, AND THE GEOGRAPHICAL DISTRIBUTION OF ANIMALS

It requires unusual comprehensiveness of view to perceive the order prevailing in the geographical distribution of animals. We should therefore not wonder that this branch of Zoology is so far

[144] *The Classification of Insects from Embryological Data* (Smithsonian Contributions to Knowledge, II, Washington, 1850).

[145] Agassiz, *Lake Superior . . .* (Boston, 1850), p. 194.

behind the other divisions of that science. Nor should we wonder at the fact that the geographical distribution of plants is so much better known than that of animals, when we consider how marked a feature the vegetable carpet which covers the surface of our globe is, when compared with the little show animals make, almost everywhere. And yet it will perhaps some day be easier to understand the relations existing between the geographical distribution of animals and the other general relations prevailing among animals, because the range of structural differences is much greater among animals than among plants. Even now some curious coincidences may be pointed out which go far to show that the geographical distribution of animals stands in direct relation to their relative standing in their respective classes, and to the order of their succession in past geological ages, and more indirectly also to their embryonic growth.

Almost every class has its tropical families, and these stand generally highest in their respective classes; or, when the contrary is the case, when they stand evidently upon a lower level, there is some prominent relation between them and the prevailing types of past ages. The class of Mammalia affords striking examples of these two kinds of connection. In the first place, the Quadrumana, which, next to Man, stand highest in their class, are all tropical animals; and it is worthy of remark that the two highest types of Anthropoid Monkeys, the Orangs of Asia and the Chimpanzees of Western Africa bear in the coloration of their skin an additional similarity to the races of Man inhabiting the same regions, the Orangs being yellowish red, as the Malays, and the Chimpanzee blackish, as the Negroes. The Pachyderms, on the contrary, stand low in their class, though chiefly tropical; but they constitute a group of animals prominent among the earliest representatives of that class in past ages. Among Chiroptera the larger frugivorous representatives are essentially tropical; the more omnivorous, on the contrary, occur everywhere. Among Carnivora, the largest, most powerful, and also highest types, the Digitigrade, prevail in the tropics, while among the Plantigrades, the most powerful, the Bears, belong to the temperate and to the arctic zone, and the lowest, the Pinnate, are marine species of the temperate and arctic seas. Among Ruminants we find the Giraffe and the Camels in the warmer zones, the others everywhere. In the class of Birds the gradation is not so obvious as in other classes, and yet the aquatic types form by far the largest representation of this class in temperate and cold regions and are almost the only ones found in the arctic, while the higher land birds prevail in the warm regions. Among Reptiles the Crocodilians are entirely tropical; the largest land Turtles are also only found in the tropics, and the aquatic representatives of this order, which are evidently inferior to their land kindred, extend much further north. The Rattlesnakes and Vipers extend further north and higher up the mountains than the Boas and the common harmless snakes. The same is true of Salamanders and Tritons. The Sharks and Skates are most diversified in the tropics. It is also within the tropics that the most brilliant diurnal Lepidoptera are found, and this is the highest order of Insects. Among Crustacea the highest order, the Brachyura, are most numerous in the torrid zone; but Dana has shown, what was not at all expected, that they nevertheless reach their highest perfection in the middle temperate regions. The Anomoura and Macroura, on the contrary, are nearly equally divided between the torrid and temperate zones; while the lower Tetradecapods are far more numerous in extra tropical latitudes than in the tropical. The Cephalopods are most diversified within the tropics; yet the Nautilus is a reminiscence of past ages. Among Gasteropods, the Stromboids belong to the tropics; but among the lamellibranchiate Acephala, the Naiades, which seem to me to stand very high in their class, have their greatest development in the fresh waters of North America. The highest Echinoderns, the Holothurians and Spatangoids are most diversified within the tropics, while Echini, Starfishes, and Ophiuræ extend to the arctics. The presence of Pentacrinus in the West Indies has undoubtedly reference to the prevalence of Crinoids in past ages. The Madrepores, the highest among the Actinoid Polypi, are entirely tropical, while the highest Halcyonoids, the Renilla, Veretillum, and Pennatula, extend to the tropics and the temperate zone.

Another interesting relation between the geographical distribution of animals and their representatives in past ages is the absence of embryonic types in the warm regions. We find in the torrid zone no true representatives of the oldest geological periods; Pentacrinus is not found before the Lias; among Cephalopods we find the Nautilus, but nothing like Orthoceras; Limulus, but nothing like Trilobites.

This study of the relations between the geographical distribution of animals and their relative standing is rendered more difficult and in many respects obscure by the circumstance that entire types, characterized by peculiar structures, are so strangely limited in their range; and yet even this shows how closely the geographical distribution of animals is connected with their structure. Why New Holland should have no Monkeys, no Carnivora, no Ruminants, no Pachyderms, no Edentata, is not to be explained; but that this is the case, every zoologist knows and is further aware that the Marsupials of that continental island represent, as it were, the other orders of Mammalia, under their special structural modifications. New Holland appears thus as a continent with the characters of an older geological age. No one can fail therefore to perceive of how great an interest for Classification will be a more extensive knowledge of the geographical distribution of animals in general and of the structural peculiarities exhibited by localized types.

SECTION XXIX

MUTUAL DEPENDENCE OF THE ANIMAL AND VEGETABLE KINGDOMS

Though it had long been known, by the experiments of de Saussure[146] that the breathing process is very different in animals and plants and that while the former inhale atmospheric air and exhale carbonic acid gas, the latter appropriate carbon and exhale oxygen, it was not until Dumas and Boussingault[147] called particularly the attention of naturalists to the subject that it was fully understood how direct the dependence is of the animal and vegetable kingdoms one upon the other in that respect, or rather how the one consumes what the other produces, and *vice versâ,* thus tending to keep the

[146] [Horace Bénédict de Saussure, 1740–1799.]

[147] Jean B. A. Dumas, "Lecon sur la statique chimique des êtres organisés," *Annales des Sciences Naturelles,* VI (2d ser., 1836), 33, "Additions . . . ," in *ibid.,* XVII (2d ser., 1842), 122; Dumas and Jean B. J. Boussingault, "Recherches sur l'engraissement des bestiaux et la formation du lait," in *ibid.,* XIX (2d ser., 1843), 351.

balance which either of them would singly disturb to a certain degree. The common agricultural practice of manuring exhibits from another side the dependence of one kingdom upon the other: the undigested particles of the food of animals return to the ground to fertilize it for fresh production. Again, the whole animal kingdom is either directly or indirectly dependent upon the vegetable kingdom for its sustenance, as the herbivorous animals afford the needful food for the carnivorous tribes. We are too far from the time when it could be supposed that Worms originated in the decay of fruits and other vegetable substances to need here repetition of what is known respecting the reproduction of these animals. Nor can it be necessary to show how preposterous the assumption would be that physical agents produced plants first, in order that from these, animals might spring forth. Who could have taught the physical agents to make the whole animal world dependent upon the vegetable kingdom?

On the contrary, such general facts as those above alluded to show more directly than any amount of special disconnected facts could do the establishment of a well-regulated order of things considered in advance; for they exhibit well-balanced conditions of existence, prepared long beforehand, such as only an intelligent being could ordain.

SECTION XXX

PARASITIC ANIMALS AND PLANTS

However independent of each other some animals may appear, there are yet many which live only in the closest connection with their fellow-creatures and which are known only as parasites upon or within them. Such are the intestinal Worms and all the vermin of the skin. Among plants the Mistletoe, Orobanche, Rafflesia, and many Orchideæ may be quoted as equally remarkable examples of parasitism.

There exists the greatest variety of parasites among animals. It would take volumes to describe them and to write their history, for their relations to the animals and plants upon which they are dependent for their existence are quite as diversified as their form and their structure.

It is important, however, to remark at the outset that these parasites do not constitute for themselves one great division of the animal kingdom. They belong, on the contrary, to all its branches; almost every class has its parasites, and in none do they represent one natural order. This fact is very significant, as it shows at once that parasitism is not based upon peculiar combinations of the leading structural features of the animal kingdom, but upon correlations of a more specific character. Nor is the degree of dependence of parasites upon other organized beings equally close. There are those which only dwell upon other animals, while others are so closely connected with them that they cannot subsist for any length of time out of the most intimate relation to the species in which they grow and multiply. Nor do these parasites live upon one class of animals; on the contrary, they are found in all of them.

Among Vertebrata there are few parasites, properly speaking. None among Mammalia. Among Birds a few species depend upon others to sit upon their eggs and hatch them, as the European Cuckoo and the North American Cowbird. Among Fishes some small Ophidiums (Fierasfers) penetrate into the cavity of the body of large Holothuriæ in which they dwell. Echeneis attach themselves to other fishes, but only temporarily. Among Articulata the number of parasites is largest. It seems to lie in the very character of this type, so remarkable for the outward display of their whole organization, to include the greatest variety of parasites. And it is really among them that we observe the most extraordinary combinations of this singular mode of existence.

Insects in general are more particularly dependent upon plants for their sustenance than herbivorous animals usually are, inasmuch as most of them are limited to particular plants for their whole life, such as the Plant-lice, the Coccus, the Gall Insects. In others the larvæ only are so limited to particular plants, while the larvæ of others still, such as the Bots, grow and undergo their development under the skin or in the intestines, or in the nasal cavities of other animals. The Ichneumons lay their eggs in the larvæ of other insects, upon which the young larvæ prey until hatched. Among perfect Insects there are those which live only in community with others, such as the Ant-Hill Insects, the Clavigers, the Cleri, and Bees. Different kinds of Ants live together, if not as parasites one upon another, at least in a kind of servitude. Other Insects live upon the bodies of warm-blooded animals, such as the Fleas and Lice, and of these the number is legion. Some Hydrachnas are parasitic upon aquatic Mollusks.

Among Crustacea there are Crabs constantly living in the shell of Mollusks, such as the Pinnotheres of the Oyster and Mussel. I have found other species upon Sea-Urchins (*Pinnotheres Melittæ,* a new species, upon *Melitta quinquefora*). The Paguri take the shells of Mollusks to protect themselves; while a vast number of Amphipods live upon Fishes, attached to their gills, upon their tongue, or upon their skin, or upon Starfishes. The *Cyamus Ceti* lives upon the Whale. Some Cirripeds are parasites upon the Whales, others upon Corals. In the family of Lernæans the females are mostly parasites upon the gills or fins or upon the body of Fishes, while the males are free.

Among Worms this mode of existence is still more frequent, and while some dwell only among Corals, entire families of others consist only of genuine parasites; but here again we find the most diversified relations; for, while some are constantly parasitic, others depend only for a certain period of their life upon other animals for their existence. The young Gordius is a free animal; it then creeps into the body of Insects and leaves them again to propagate; the young Distoma lives free in the water as Cercaria and spends the remainder of its life in other animals; the Tænia, on the contrary, is a parasite through life, and only its eggs pass from one animal into the other. But what is most extraordinary in this, as in many other intestinal Worms, is the fact that while they undergo their first transformations in some kind of animals they do not reach their complete development until they pass into the body of another higher type, being swallowed up by this while in the body of their first host. Such is the case with many Filariæ, the Tæniæ and Bothrocephali. These at first inhabit lower Fishes, and these Fishes being swallowed by Sharks or Water Birds, or Mice with their Worms being eaten up by Cats, the parasites living in them undergo their final transformation in the latter. Many Worms undertake extensive migrations through the bodies of other animals before they reach the proper place for their final development.

Among Mollusks parasites are very few, if any can properly be called true parasites, as the males of some Cephalopods living upon their own females; as the Gasteropods growing buried in Corals, and the Lithodomus and a variety of Arcas found in Corals. Among Radiata there are no parasites, properly speaking; some of them only attaching themselves by preference to certain plants, while the young of others remain connected with their parent, as in all Corals, and even among Crinoids, as in the Comatula of Charleston.

In all these different cases the chances that physical agents may have a share in producing such animals are still less than in the cases of independent animals, for here we have superadded to the very existence of these beings all the complicated circumstances of their peculiar mode of existence and their various connections with other animals. Now if it can already be shown from the mere connections of independent animals that external circumstances cannot be the cause of their existence, how much less could such an origin be ascribed to parasites! It is true they have been supposed to originate in the body of the animals upon which they live. What then of those who enter the body of other animals at a somewhat advanced stage of growth, as the Gordius? Is it a freak of his? Or what of those which only live upon other animals, such as lice; are they the product of the skin? Or what of those which have to pass from the body of a lower into that of a higher animal, to undergo their final metamorphosis and in which this succession is normal? Was such an arrangement devised by the first animal, or imposed upon the first by the second, or devised by physical agents for the two? Or what of those in which the females only are parasites? Had the two sexes a different origin? Did perhaps the males and females originate in different ways?

I am at a loss to conceive how the origin of parasites can be ascribed to physical causes, unless indeed animals themselves be considered as physical causes with reference to the parasites they nourish; and if so, why can they not get rid of them, as well as produce them, for it cannot be supposed, that all this is not done consciously, when parasites bear such close structural relations to the various types to which they belong?

The existence of parasitic animals belonging to so many different types of the animal as well as the vegetable kingdom is a fact of deep meaning which Man himself cannot too earnestly consider, and, while he may marvel at the fact, take it as a warning for himself with reference to his boasted and yet legitimate independence. All relations in nature are regulated by a superior wisdom. May we only learn in the end to conform, within the limits of our own sphere, to the laws assigned to each race!

SECTION XXXI

COMBINATIONS IN TIME AND SPACE OF VARIOUS KINDS OF RELATIONS AMONG ANIMALS

It must occur to every reflecting mind, that the mutual relation and respective parallelism of so many structural, embryonic, geological, and geographical characteristics of the animal kingdom are the most conclusive proof that they were ordained by a reflective mind, while they present at the same time the side of nature most accessible to our intelligence, when seeking to penetrate the relations between finite beings and th cause of their existence.

The phenomena of the inorganic world are all simple, when compared to those of the organic world. There is not one of the great physical agents, electricity, magnetism, heat, light, or chemical affinity, which exhibits in its sphere as complicated phenomena as the simplest organized beings; and we need not look for the highest among the latter to find them presenting the same physical phenomena as are manifested in the material world, besides those which are exclusively peculiar to them. When then organized beings include everything the material world contains and a great deal more that is peculiarly their own, how could they be produced by physical causes, and how can the physicists, acquainted with the laws of the material world and who acknowledge that these laws must have been established at the beginning, overlook that *à fortiori* the more complicated laws which regulate the organic world, of the existence of which there is no trace for a long period upon the surface of the earth, must have been established later and successively at the time of the creation of the successive types of animals and plants?

Thus far we have been considering chiefly the contrasts existing between the organic and inorganic worlds. At this stage of our investigation it may not be out of place to take a glance at some of the coincidences which may be traced between them, especially as they afford direct evidence that the physical world has been ordained in conformity with laws which obtain also among living beings, and disclose in both spheres equally plainly the workings of a reflective mind. It is well known that the arrangement of the leaves in plants[148] may be expressed by very simple series of fractions, all of which are gradual approximations to, or the natural means between $\frac{1}{2}$ or $\frac{1}{3}$, which two fractions are themselves the maximum and the minimum divergence between two single successive leaves. The normal series of fractions which expresses the various combinations most frequently observed among the leaves of plants is as follows: $\frac{1}{2}$, $\frac{1}{3}$, $\frac{2}{5}$, $\frac{3}{8}$, $\frac{5}{13}$, $\frac{8}{21}$, $\frac{13}{34}$, $\frac{21}{55}$, etc. Now upon comparing this arrangement of the leaves in plants with the revolutions of the members of our solar system, Peirce has discovered the most perfect identity between the fundamental laws which regulate both, as may be at once seen by the following diagram, in which the first column gives the names of the planets, the second column indicates the actual time of revolution of the successive planets, expressed in days; the third column, the successive times of revolution of the planets, which are derived from the hypothesis that each time of revolution should have a ratio to those upon each side of it, which shall be one of the ratios of the law of phyllotaxis; and the fourth column, finally, gives the normal series of fractions expressing the law of the phyllotaxis.[149]

Neptune,	60,129	62,000		
Uranus,	30,687	31,000		$\frac{1}{2}$
Saturn,	10,759	10,333		$\frac{1}{3}$
Jupiter,	4,333	4,133		$\frac{2}{5}$
Asteroids,	1,200 to 2,000	1,550		$\frac{3}{8}$
Mars,	687	596		$\frac{5}{13}$
Earth,	365	366	$\frac{8}{13}$	$\frac{8}{21}$
Venus,	225	227	$\frac{13}{21}$	
Mercury,	88	87		$\frac{13}{34}$

[148] Johann Wolfgang von Göthe, *Zur Naturwissenschaft überhaupt, besonders zur Morphologie* (2 vols., Stuttgart, 1817–1824), and *Oeuvres d'histoire naturelle, comprenant divers mémoires d'anatomie comparée, de botanique et de géologie* . . . (Paris, 1837); Augustin P. de Candolle, *Organographie végétale* (2 vols., Paris, 1827); Braun, *Das Individuum der Pflanze.*

[149] [Agassiz tried to interest Americans in this concept, an idea typical of German speculative biology and one that he had been much impressed with since his student days at the University of Munich. See Asa Gray, "On the Composition of the Plant by Phytons, and Some Applications of Phyllotaxis," *Proceedings,* AAAS, II (1850), 438–444, and Benjamin Peirce, "Mathematical Investigations of the Fractions Which Occur in Phyllotaxis," in *ibid.,* 444–447. Gray was never entirely convinced of the validity of this ideal conception. He subsequently encouraged Chauncey Wright to examine the problem of leaf arrangement, with the result that such facts were shown to be understandable in terms of the principle of natural selection.]

In this series the Earth forms a break; but this apparent irregularity admits of an easy explanation. The fractions $\frac{1}{2}, \frac{1}{3}, \frac{2}{5}, \frac{3}{8}, \frac{5}{13}, \frac{8}{21}, \frac{13}{24}$, etc., as expressing the position of successive leaves upon an axis, by the short way of ascent along the spiral, are identical as far as their meaning is concerned with the fractions expressing these same positions by the long way, namely, $\frac{1}{2}, \frac{2}{3}, \frac{3}{5}, \frac{5}{8}, \frac{8}{13}, \frac{13}{21}, \frac{21}{34}$, etc.

Let us therefore repeat our diagram in another form, the third column giving the theoretical time of revolution.

Neptune,	$\frac{1}{1}$	62,000	60,129
"	$\frac{1}{1}$	62,000	——
Uranus,	$\frac{1}{2}$	31,000	30,687
"	$\frac{1}{2}$	15,500	——
Saturn,	$\frac{2}{3}$	10,333	10,759
"	$\frac{2}{3}$	6,889	——
Jupiter,	$\frac{3}{5}$	4,133	4,333
"	$\frac{3}{5}$	2,480	——
Asteroids,	$\frac{5}{8}$	1,550	1,200
"	$\frac{5}{8}$	968	——
Mars,	$\frac{8}{13}$	596	687
Earth,	$\frac{8}{13}$	366	365
Venus,	$\frac{13}{21}$	227	225
"	$\frac{13}{21}$	140	——
Mercury,	$\frac{21}{34}$	87	88

It appears from this table that two intervals usually elapse between two successive planets, so that the normal order of actual fractions is $\frac{1}{2}, \frac{1}{3}, \frac{2}{5}, \frac{3}{8}, \frac{5}{13}$, etc., or the fractions by the short way in phyllotaxis, from which, however, the Earth is excluded, while it forms a member of the series by the long way. The explanation of this, suggested by Peirce, is that although the tendency to set off a planet is not sufficient at the end of a single interval, it becomes so strong near the end of the second interval, that the planet is found exterior to the limit of this second interval. Thus, Uranus is rather too far from the Sun relatively to Neptune, Saturn relatively to Uranus, and Jupiter relatively to Saturn; and the planets thus formed engross too large a proportionate share of material, and this is especially the case with Jupiter. Hence, when we come to the Asteroids, the disposition is so strong at the end of a single interval, that the outer Asteroid is but just within this interval, and the whole material of the Asteroids is dispersed in separate masses over a wide space, instead of being concentrated into a single planet. A consequence of this dispersion of the forming agents is that a small proportionate material is absorbed into the Asteroids. Hence, Mars is ready for formation so far exterior to its true place, that when the next interval elapses the residual force becomes strong enough to form the Earth, after which the normal law is resumed without any further disturbance. Under this law there can be no planet exterior to Neptune, but there may be one interior to Mercury.

Let us now look back upon some of the leading features alluded to before, omitting the simpler relations of organized beings to the world around, or those of individuals to individuals, to consider only the different parallel series we have been comparing when showing that in their respective great types the phenomena of animal life correspond to one another, whether we compare their rank as determined by structural complication with the phases of their growth, or with their succession in past geological ages; whether we compare this succession with their embryonic growth, or all these different relations with each other and with the geographical distribution of animals upon earth. The same series everywhere! These facts are true of all the great divisions of the animal kingdom, so far as we have pursued the investigation; and though, for want of materials, the train of evidence is incomplete in some instances, yet we have proof enough for the establishment of this law of a universal correspondence in all the leading features which binds all organized beings of all times into one great system, intellectually and intelligibly linked together, even where some links of the chain are missing. It requires considerabe familiarity with the subject even to keep in mind the evidence, for, though yet imperfectly understood, it is the most brilliant result of the combined intellectual efforts of hundreds of investigators during half a century. The connection, however, between the facts, it is easily seen, is only intellectual; and implies therefore the agency of Intellect as its first cause.[150]

And if the power of thinking connectedly is the privilege of cultivated minds only; if the power of combining different thoughts and of drawing from them new thoughts is a still rarer privilege of a few superior minds; if the ability to trace simultaneously several trains of thought is such an extraordinary gift, that the few cases in which evidence of this kind has been presented have become a

[150] Agassiz, "Contemplations of God in the Cosmos," *Christian Examiner,* L (1851), 1–17.

matter of historical record (Cæsar dictating several letters at the same time), though they exhibit only the capacity of passing rapidly, in quick succession, from one topic to another, while keeping the connecting thread of several parallel thoughts: if all this is only possible for the highest intellectual powers, shall we by any false argumentation allow ourselves to deny the intervention of a Supreme Intellect in calling into existence combinations in nature, by the side of which all human conceptions are child's play?

If I have succeeded, even very imperfectly, in showing that the various relations observed between animals and the physical world, as well as between themselves, exhibit thought, it follows that the whole has an Intelligent Author; and it may not be out of place to attempt to point out, as far as possible, the difference there may be between Divine thinking and human thought.

Taking nature as exhibiting thought for my guide, it appears to me that while human thought is consecutive, Divine thought is simultaneous, embracing at the same time and forever, in the past, the present, and the future, the most diversified relations among hundreds of thousands of organized beings, each of which may present complications again, which, to study and understand even imperfectly, as for instance, Man himself, Mankind has already spent thousands of years. And yet, all this has been done by one Mind, must be the work of one Mind only, of Him before whom Man can only bow in grateful acknowledgment of the prerogatives he is allowed to enjoy in this world, not to speak of the promises of a future life.

I have intentionally dismissed many points in my argument with mere questions, in order not to extend unduly a discussion which is after all only accessory to the plan of my work. I have felt justified in doing so because, from the point of view under which my subject is treated, those questions find a natural solution which must present itself to every reader. We know what the intellect of Man may originate, we know its creative power, its power of combination, of foresight, of analysis, of concentration; we are, therefore, prepared to recognize a similar action emanating from a Supreme Intelligence to a boundless extent. We need therefore not even attempt to show that such an Intellect may have originated all the Universe contains; it is enough to demonstrate that the constitution of the physical world and, more particularly, the organization of living beings in their connection with the physical world, prove in general the existence of a Supreme Being as the Author of all things. The task of science is rather to investigate what has been done, to inquire if possible how it has been done, than to ask what is possible for the Deity, as we can know that only by what actually exists. To attack such a position, those who would deny the intervention in nature of a creative mind must show that the cause to which they refer the origin of finite beings is by its nature a possible cause, which cannot be denied of a being endowed with the attributes we recognize in God. Our task is therefore completed as soon as we have proved His existence. It would nevertheless be highly desirable that every naturalist who has arrived at similar conclusions should go over the subject anew from his point of view and with particular reference to the special field of his investigations; for so only can the whole evidence be brought out.

I foresee already that some of the most striking illustrations may be drawn from the morphology of the vegetable kingdom, especially from the characteristic succession and systematical combination of different kinds of leaves in the formation of the foliage and the flowers of so many plants, all of which end their development by the production of an endless variety of fruits. The inorganic world, considered in the same light, would not fail to exhibit also unexpected evidence of thought, in the character of the laws regulating the chemical combinations, the action of physical forces, the universal attraction, etc., etc. Even the history of human culture ought to be investigated from this point of view. But I must leave it to abler hands to discuss such topics.

SECTION XXXII

RECAPITULATION

In recapitulating the preceding statements we may present the following conclusions: —

1st.[151] The connection of all these known features of nature into

[151] The numbers inscribed here correspond to the preceding sections, in the same order, so that the reader may at once refer back to the evidence, when needed.

one system exhibits thought, the most comprehensive thought, in limits transcending the highest wonted powers of man.

2d. The simultaneous existence of the most diversified types under identical circumstances exhibits thought, the ability to adapt a great variety of structures to the most uniform conditions.

3d. The repetition of similar types, under the most diversified circumstances, shows an immaterial connection between them; it exhibits thought, proving directly how completely the Creative Mind is independent of the influence of a material world.

4th. The unity of plan in otherwise highly diversified types of animals exhibits thought; it exhibits more immediately premeditation, for no plan could embrace such a diversity of beings, called into existence at such long intervals of time, unless it had been framed in the beginning with immediate reference to the end.

5th. The correspondence, now generally known as special homologies, in the details of structure in animals otherwise entirely disconnected, down to the most minute peculiarities, exhibits thought, and more immediately the power of expressing a general proposition in an indefinite number of ways, equally complete in themselves, though differing in all their details.

6th. The various degrees and different kinds of relationship among animals which can have no genealogical connection, exhibit thought, the power of combining different categories into a permanent, harmonious whole, even though the material basis of this harmony be ever changing.

7th. The simultaneous existence, in the earliest geological periods in which animals existed at all, of representatives of all the great types of the animal kingdom exhibits most especially thought, considerate thought, combining power, premeditation, prescience, omniscience.

8th. The gradation based upon complications of structure, which may be traced among animals built upon the same plan, exhibits thought, and especially the power of distributing harmoniously unequal gifts.

9th. The distribution of some types over the most extensive range of the surface of the globe, while others are limited to particular geographical areas, and the various combinations of these types into zoological provinces of uneequal extent exhibit thought, a close control in the distribution of the earth's surface among its inhabitants.

10th. The identity of structure of these types, notwithstanding their wide geographical distribution, exhibits thought, that deep thought which, the more it is scrutinized, seems the less capable of being exhausted, though its meaning at the surface appears at once plain and intelligible to every one.

11th. The community of structure in certain respects of animals otherwise entirely different, but living within the same geographical area, exhibits thought, and more particularly the power of adapting most diversified types with peculiar structures to either identical or to different conditions of existence.

12th. The connection, by series, of special structures observed in animals widely scattered over the surface of the globe exhibits thought, unlimited comprehension, and more directly omnipresence of mind and also prescience, as far as such series extend through a succession of geological ages.

13th. The relation there is between the size of animals and their structure and form exhibits thought; it shows that in nature the quantitative differences are as fixedly determined as the qualitative ones.

14th. The independence in the size of animals of the mediums in which they live exhibits thought, in establishing such close connection between elements so influential in themselves and organized beings so little affected by the nature of these elements.

15th. The permanence of specific peculiarities under every variety of external influences during each geological period and under the present state of things upon earth exhibits thought: it shows also that limitation in time is an essential element of all finite beings, while eternity is an attribute of the Deity only.

16th. The definite relations in which animals stand to the surrounding world exhibit thought; for all animals living together stand respectively, on account of their very differences, in different relations to identical conditions of existence, in a manner which implies a considerate adaptation of their varied organization to these uniform conditions.

17th. The relations in which individuals of the same species stand to one another exhibit thought and go far to prove the existence in all living beings of an immaterial, imperishable principle, similar to that which is generally conceded to man only.

18th. The limitation of the range of changes which animals undergo during their growth exhibits thought; it shows most strikingly the independence of these changes of external influences and the necessity that they should be determined by a power superior to these influences.

19th. The unequal limitation in the average duration of the life of individuals in different species of animals exhibits thought; for, however uniform or however diversified the conditions of existence may be under which animals live together, the average duration of life in different species is unequally limited. It points therefore at a knowledge of time and space and of the value of time, since the phases of life of different animals are apportioned according to the part they have to perform upon the stage of the world.

20th. The return to a definite norm of animals which multiply in various ways exhibits thought. It shows how wide a cycle of modulations may be included in the same conception, without yet departing from a norm expressed more directly in other combinations.

21st. The order of succession of the different types of animals and plants characteristic of the different geological epochs exhibits thought. It shows that while the material world is identical in itself in all ages ever different types of organized beings are called into existence in successive periods.

22d. The localization of some types of animals upon the same points of the surface of the globe, during several successive geological periods, exhibits thought, consecutive thought; the operations of a mind acting in conformity with a plan laid out beforehand and sustained for a long period.

23d. The limitation of closely allied species to different geological periods exhibits thought; it exhibits the power of sustaining nice distinctions, notwithstanding the interposition of great disturbances by physical revolutions.

24th. The parallelism between the order of succession of animals and plants in geological times and the gradation among their living representatives exhibit thought; consecutive thought, superintending the whole development of nature from beginning to end, and disclosing throughout a gradual progress, ending with the introduction of man at the head of the animal creation.

25th. The parallelism between the order of succession of animals in geological times and the changes their living representatives undergo during their embryological growth exhibits thought; the repetition of the same train of thoughts in the phases of growth of living animals and the successive appearance of their representatives in past ages.

26th. The combination in many extinct types of characters which, in later ages, appear disconnected in different types exhibits thought, prophetic thought, foresight; combinations of thought preceding their manifestation in living forms.

27th. The parallelism between the gradation among animals and the changes they undergo during their growth exhibits thought, as it discloses everywhere the most intimate connection between essential features of animals which have no necessary physical relation, and can therefore not be understood otherwise than as established by a thinking being.

28th. The relations existing between these different series and the geographical distribution of animals exhibit thought; they show the omnipresence of the Creator.

29th. The mutual dependence of the animal and vegetable kingdoms for their maintenance exhibits thought; it displays the care with which all conditions of existence necessary to the maintenance of organized beings have been balanced.

30th. The dependence of some animals upon others or upon plants for their existence exhibits thought; it shows to what degree the most complicated combinations of structure and adaptation can be rendered independent of the physical conditions which surround them.

We may sum up the results of this discussion, up to this point, in still fewer words: —

All organized beings exhibit in themselves all those categories of structure and of existence upon which a natural system may be founded, in such a manner that, in tracing it, the human mind is only translating into human language the Divine thoughts expressed in nature in living realities.

All these beings do not exist in consequence of the continued agency of physical causes, but have made their successive appearance upon earth by the immediate intervention of the Creator. As proof I may sum up my argument in the following manner:

The products of what are commonly called physical agents are everywhere the same (that is, upon the whole surface of the globe) and have always been the same (that is, during all geological periods); while organized beings are everywhere different and have differed in all ages. Between two such series of phenomena there can be no causal or genetic connection.

31st. The combination in time and space of all these thoughtful conceptions exhibits not only thought, it shows also premeditation, power, wisdom, greatness, prescience, omniscience, providence. In one word, all these facts in their natural connection proclaim aloud the One God, whom man may know, adore, and love; and Natural History must in good time become the analysis of the thoughts of the Creator of the Universe, as manifested in the animal and vegetable kingdoms, as well as in the inorganic world.

It may appear strange that I should have included the preceding disquisition under the title of an "Essay on Classification." Yet it has been done deliberately. In the beginning of this chapter I have already stated that Classification seems to me to rest upon too narrow a foundation when it is chiefly based upon structure. Animals are linked together as closely by their mode of development, by their relative standing in their respective classes, by the order in which they have made their appearance upon earth, by their geographical distribution, and generally by their connection with the world in which they live, as by their anatomy. All these relations should therefore be fully expressed in a natural classification; and though structure furnishes the most direct indication of some of these relations, always appreciable under every circumstance, other considerations should not be neglected which may complete our insight into the general plan of creation.

In characterizing the great branches of the animal kingdom it is not enough to indicate the plan of their structure in all its peculiarities; there are possibilities of execution which are at once suggested to the exclusion of others, and which should also be considered and so fully analyzed, that the various modes in which such a plan may be carried out shall at once be made apparent. The range and character of the general homologies of each type should also be illustrated, as well as the general conditions of existence of its representatives. In characterizing classes it ought to be shown why such groups constitute a class and not merely an order or a family; and to do this satisfactorily it is indispensable to trace the special homologies of all the systems of organs which are developed in them. It is not less important to ascertain the foundation of all the subordinate divisions of each class; to know how they differ, what constitutes orders, what families, what genera, and upon what characteristics species are based in every natural division. This we shall examine in the next chapter.

CHAPTER II

LEADING GROUPS OF THE EXISTING SYSTEMS OF ANIMALS

SECTION I

GREAT TYPES OR BRANCHES OF THE ANIMAL KINGDOM

The use of the terms types, classes, orders, families, genera, and species in the systems of Zoology and Botany is so universal, that it would be natural to suppose that their meaning and extent are well determined and generally understood; but this is so far from being the case that it may on the contrary be said that there is no subject in Natural History respecting which there exists more uncertainty or a greater want of precision. Indeed, I have failed to find anywhere a definition of the character of most of the more comprehensive of these divisions, while the current views respecting genera and species are very conflicting. Under these circumstances it has appeared to me particularly desirable to inquire into the foundation of these distinctions, and to ascertain if possible how far they have a real existence. And, while I hope the results of this inquiry may be welcome and satisfactory, I am free to confess that it has cost me years of labor to arrive at a clear conception of their true character.

It is such a universal fact in every sphere of intellectual activity that practice anticipates theory, that no philosopher should be surprised to find that zoologists have adopted instinctively natural groups in the animal and vegetable kingdoms, even before the question of the character and of the very existence of such groups in nature was raised. Did not nations speak, understand, and write Greek, Latin, German, and Sanscrit, before it was even suspected that these languages and so many others were kindred? Did not painters produce wonders with colors before the nature of light was understood? Had not men been thinking about themselves and the world before logic and metaphysics were taught in schools? Why then should not observers of nature have appreciated rightly the relationship between animals or plants before getting a scientific clue to the classifications they were led to adopt as practical?

Such considerations, above all others, have guided and encouraged me while I was seeking for the meaning of all these systems, so different one from the other in their details, and yet so similar in some of their general features. The history of our science shows how early some of the principles, which obtain to this day, have been acknowledged by all reflecting naturalists. Aristotle, for instance, already knew the principal differences which distinguish Vertebrata from all other animals, and his distinction of *Enaima* and *Anaima* corresponds exactly to that of *Vertebrata* and *Invertebrata* of Lamarck, or to that of *Flesh-* and *Gut-Animals* of Oken, or to that of *Myeloneura* and *Ganglioneura* of Ehrenberg; and one who is at all familiar with the progress of science at different periods can but smile at the claims to novelty or originality so frequently brought forward for views long before current among men. Here, for instance, is one and the same fact presented in different aspects; first by Aristotle with reference to the character of the formative fluid, next by Lamarck with reference to the general frame — for I will do Lamarck the justice to believe that he did not unite the Invertebrata simply because they have no skeleton but because of that something, which even Professor Owen fails to express and which yet exists, the one cavity of the body in Invertebrata containing all organs, whilst Vertebrata have one distinct cavity for the centres of the nervous system and another for the organs of the vegetative life. This acknowledgment is due to Lamarck as truly as it would be due to Aristotle not to accuse him of having denied the Invertebrata any fluid answering the office of the blood, though he calls them *Anaima;* for he knew nearly as well as we now know that there moves a nutritive fluid in their body, though that information is generally denied him because he had no correct knowledge of the circulation of the blood.

Again, when Oken speaks of Flesh-Animals he does not mean that Vertebrates consist of nothing but flesh or that the Invertebrates have no muscular fibres; but he brings prominently before us the presence, in the former, of those masses forming mainly the bulk of the body, which consist of flesh and bones as well as blood and nerves and constitute another of the leading features distinguishing Vertebrata and Invertebrata. Ehrenberg presents the same relations between the same beings as expressed by their nervous system. If we now take the expressions of Aristotle, Lamarck, Oken, and Ehrenberg together, have we not, as characteristic of their systems, the very words by which every one distinguishes the most prominent features of the body of the higher animals, when speaking of blood relations, of blood and bones, or of having flesh and nerve?

Neither of these observers has probably been conscious of the identity of his classification with that of his predecessors; nor indeed should we consider either of them as superfluous, inasmuch as it makes prominent, features more or less different from those insisted upon by the others; nor ought any one to suppose that with all of them the field is exhausted and that there is no more room for new systems upon that very first distinction among animals.[1] As long as men inquire, they will have opportunities to know more upon these topics than those who have gone before them, so inexhaustibly rich is nature in the innermost diversity of her treasures of beauty, order, and intelligence.

So, instead of discarding all the systems which have thus far had little or no influence upon the progress of science, either because they are based upon principles not generally acknowledged or considered worthy of confidence, I have carefully studied them with the view of ascertaining whatever there may be true in them, from the stand-point from which their authors have considered the animal kingdom; and I own that I have often derived more information from such a careful consideration than I had at first expected.

It was not indeed by a lucky hit, nor by one of those unexpected apparitions which, like a revelation, suddenly break upon us and render at once clear and comprehensible what had been dark and

[1] By way of an example I would mention the mode of reproduction. The formation of the egg in Vertebrata; its origin in all of them in a more or less complicated Graafian vesicle, in which it is nursed; the formation and development of the embryo up to a certain period, etc., etc., are so completely different from what is observed in any of the Invertebrata, that the animal kingdom, classified according to these facts, would again be divided into two great groups corresponding to the *Vertebrata* and *Invertebrata* of Lamarck, or the *Flesh-* and *Gut-Animals* of Oken, or the *Enaima* and *Anaima* of Aristotle, etc.

almost inaccessible before, that I came to understand the meaning of those divisions called types, classes, orders, families, genera, and species, so long admitted in Natural History as the basis of every system, and yet so generally considered as mere artificial devices to facilitate our studies. For years I had been laboring under the impression that they are founded in nature, before I succeeded in finding out upon what principle they were really based. I soon perceived, however, that the greatest obstacle in the way of ascertaining their true significance lay in the discrepancies among different authors in their use and application of these terms. Different naturalists do not call by the same name groups of the same kind and the same extent: some call genera what others call sub-genera; others call tribes or even families what are called genera by others; even the names of tribe and family have been applied by some to what others call sub-genera; some have called families what others have called orders; some consider as orders what others have considered as classes; and there are even genera of some authors which are considered as classes by others. Finally, in the number and limitation of these classes, as well as in the manner in which they are grouped together under general heads, there is found the same diversity of opinion. It is nevertheless possible that under these manifold names, so differently applied, groups may be designated which may be natural, even if their true relation to one another have thus far escaped our attention.

It is already certain that most if not all investigators agree in the limitation, of some groups at least, under whatever name they may call them, and however much they would blame one another for calling them so, or otherwise. I can, therefore, no longer doubt that the controversy would be limited to definite questions, if naturalists could only be led to an agreement respecting the real nature of each kind of group. I am satisfied indeed that the most insuperable obstacle to any exact appreciation of this subject lies in the fact that all naturalists without exception consider these divisions, under whatever name they may designate them, as strictly subordinate one to the other, in such a manner, that their difference is only dependent upon their extent; the class being considered as the more comprehensive division, the order as the next extensive, the family as more limited, the genus as still more limited, and the species as the ultimate limitation in a natural arrangement of living beings; so that all these groups would differ only by the quantity of their characters and not by the quality, as if the elements of structure in animals were all of the same kind; as if the form, for instance, was an organic element of the same kind as the complication of structure, and as if the degree of complication implied necessarily one plan of structure to the exclusion of another. I trust I shall presently be able to show that it is to a neglect of these considerations that we must ascribe the slow progress which has been made in the philosophy of classification.

Were it possible to show that all these groups do not differ in quantity and are not merely divisions of a wider or more limited range, but are based upon different categories of characters, genera would be called genera by all, whether they differ much or little one from the other, and so would families be called families, orders be called orders, etc. Could species, for instance, be based upon absolute size, genera upon the structure of some external parts of the body, families upon the form of the body, orders upon the similarity of the internal structure or the like, it is plain that there could not be two opinions respecting these groups in any class of the animal kingdom. But as the problem is not so simple in nature, it was not until after the most extensive investigations, that I obtained the clue to guide me through this labyrinth. I knew, for instance, that though naturalists have been disputing and are still disputing about species and genera, they all distinguished the things themselves in pretty much the same manner. What A would call a species, B called only a variety or a race; but then B might call a sub-genus the very same aggregate of individuals which A called a species; or what A called a genus was considered by B as a family or an order. Now it was this something called no matter how for which I tried to find out characters which would lead all to call it by the same name; thus limiting the practical difficulty in the application of the name to a question of accuracy in the observations, and no longer allowing it to be an eternal contest about mere nomenclature.

At this stage of my investigation it struck me that the character of the writings of eminent naturalists might throw some light upon the subject itself. There are authors, and among them some of the most celebrated contributors to our knowledge in Natural History, who never busied themselves with classification, or paid only a passing notice to this subject, whilst they are by universal consent considered as the most successful biographers of species; such are Buffon, Reaumur, Roesel, Trembley, Smeathman, the two Hubers, Bewick, Wilson,[2] Audubon, Naumann, etc. Others have applied themselves almost exclusively to the study of genera. Latreille[3] is the most prominent zoologist of this stamp; whilst Linnæus and Jussieu[4] stand highest among botanists for their characteristics of genera, or at least for their early successful attempts at tracing the natural limits of genera. Botanists have thus far been more successful than zoologists in characterizing natural families, though Cuvier and Latreille have done a great deal in that same direction in Zoology, whilst Linnæus was the first to introduce orders in the classification of animals. As to the higher groups, such as classes and types, and even the orders, we find again Cuvier leading the procession, in which all the naturalists of this century have followed.

Now let us inquire what these men have done in particular to distinguish themselves especially, either as biographers of species or as characterizers of genera, of families, of orders, of classes, and of types. And should it appear that in each case they have been considering their subject from some particular point of view, it strikes me that what has been acknowledged unconsciously as constituting the particular eminence or distinction of these men might very properly be proclaimed, with grateful consciousness of their services, as the characteristic of that kind of group which each of them has most successfully illustrated; and I hope every unprejudiced naturalist will agree with me in this respect.

As to the highest divisions of the animal kingdom, first introduced by Cuvier under the name of *embranchements* (and which we may well render by the good old English word *branch*) he tells us himself that they are founded upon distinct plans of structure, cast, as it were, into distinct moulds or forms.[5] Now there can certainly be

[2] [August J. von Rosenhof Roesel, 1705–1789; Abraham Trembley, 1710–1784; Henry M. Smeathman, 1750–1787; François Huber, 1750–1831; Pierre Huber, 1777–1840; Thomas Bewick, 1753–1828; Alexander Wilson, 1766–1813.]

[3] [Pierre André Latreille, 1762–1833.]

[4] [Antoine Laurent de Jussieu, 1748–1836.]

[5] It would lead me too far were I to consider here the characteristics of the different kingdoms of Nature. I may, however, refer to the work of Isidore Geoffroy St.-Hillaire, *Histoire naturelle générale . . .*, who has discussed this subject recently, though I must object to the admission of a distinct kingdom for Man alone.

no reason why we should not all agree to designate as types or branches all such great divisions of the animal kingdom as are constituted upon a special plan,[6] if we should find practically that such groups may be traced in nature. Those who may not see them may deny their existence; those who recognize them may vary in their estimation of their natural limits; but all can, for the greatest benefit of science, agree to call any group which seems to them to be founded upon a special plan of structure a type or branch of the animal kingdom; and if there are still differences of opinion among naturalists respecting their limits, let the discussion upon this point be carried on with the understanding that types are to be characterized by different plans of structure, and not by special anatomical peculiarities. Let us avoid confounding the idea of plan with that of complication of structure, even though Cuvier himself has made this mistake here and there in his classification.

The best evidence I can produce that the idea of distinct plans of structure is the true pivot upon which the natural limitation of the branches of the animal kingdom is ultimately to turn lies in the fact that every great improvement, acknowledged by all as such, which these primary divisions have undergone has consisted in the removal from among each of such groups as had been placed with them from other considerations than those of a peculiar plan, or in consequence of a want of information respecting their true plan of structure. Let us examine this point within limits no longer controvertible. Neither Infusoria nor Intestinal Worms are any longer arranged by competent naturalists among Radiata. Why they have been removed may be considered elsewhere; but it was certainly not because they

[6] It is almost superfluous for me to mention here that the terms plan, ways and means, or manner in which a plan is carried out, complication of structure, form, details of structure, ultimate structure, relations of individuals, frequently used in the following pages, are taken in a somewhat different sense from their usual meaning, as is always necessary when new views are introduced in a science, and the adoption of old expressions, in a somewhat modified sense, is found preferable to framing new ones. I trust the value of the following discussion will be appreciated by its intrinsic merit, tested with a willingness to understand what has been my aim, and not altogether by the relative degree of precision and clearness with which I may have expressed myself, as it is almost impossible in a first attempt of this kind to seize at once upon the form best adapted to carry conviction. I wish also to be understood as expressing my views more immediately with reference to the animal kingdom, as I do not feel quite competent to extend the inquiry and the discussion to the vegetable kingdom, though I have occasionally alluded to it as far as my information would permit.

were supposed to agree in the plan of their structure with the true Radiata that Cuvier placed them in that division, but simply because he allowed himself to depart from his own principle and to add another consideration, besides the plan of structure, as characteristic of Radiata — the supposed absence of a nervous system and the great simplicity of structure of these animals; — as if simplicity of execution had any necessary connection with the plan of structure. Another remarkable instance of the generally approved removal of a class from one of the types of Cuvier to another was the transfer of the Cirripeds from among the Mollusks to the branch of Articulata. Imperfect knowledge of the plan of structure of these animals was here the cause of the mistake, which was corrected without any opposition as soon as they became better known.

From a comparison of what is stated here respecting the different plans of structure characteristic of the primary divisions of the animal kingdom with what I have to say below about classes and orders, it will appear more fully, that it is important to make a distinction between the plan of a structure and the manner in which that plan is carried out, or the degrees of its complication and its relative perfection or simplicity. But even after it is understood that the plan of structure should be the leading characteristic of these primary groups, it does not yet follow, without further examination, that the four great branches of the animal kingdom, first distinguished by Cuvier, are to be considered as the primary divisions which Nature points out as fundamental. It will still be necessary, by a careful and thorough investigation of the subject, to ascertain what these primary groups are; but we shall have gained one point with reference to our systems — that whatever these primary groups, founded upon different plans, which exist in nature may be, when they are once defined, or whilst they are admitted as the temporary expression of our present knowledge, they should be called the branches of the animal kingdom, whether they be the Vertebrata, Articulata, Mollusca, and Radiata of Cuvier, or the Artiozoaria, Actinozoaria, and Amorphozoaria of Blainville, or the Vertebrata and Invertebrata of Lamarck. The special inquiry into this point must be left for a special paper. I will only add that I am daily more satisfied that in their general outlines the primary divisions of Cuvier are true to nature, and that never did a naturalist exhibit a clearer and deeper insight into the most general relations of animals than Cuvier when he perceived, not only that these primary groups are founded upon differences in the plan of their structure, but also how they are essentially related to one another.

Though the term type is generally employed to designate the great fundamental divisions of the animal kingdom, I shall not use it in future, but prefer for it the term branch of the animal kingdom, because the term type is employed in too many different acceptations, and quite as commonly to designate any group of any kind or any peculiar modification of structure stamped with a distinct and marked character as to designate the primary divisions of the animal kingdom. We speak, for instance, of specific types, generic types, family types, ordinal types, classic types, and also of a typical structure. The use of the word type in this sense is so frequent on almost every page of our systematic works, in Zoology and in treatises of Comparative Anatomy, that it seems to me desirable, in order to avoid every possible equivocation in the designation of the most important great primary divisions among animals, to call them branches of the animal kingdom, rather than types.

That, however, our systems are more true to nature than they are often supposed to be seems to me to be proved by the gradual approximation of scientific men to each other in their results and in the forms by which they express those results. The idea which lies at the foundation of the great primary divisions of the animal kingdom is the most general conception possible in connection with the plan of a definite creation; these divisions are therefore the most comprehensive of all and properly take the lead in a natural classification, as representing the first and broadest relations of the different natural groups of the animal kingdom, the general formula which they each obey. What we call branches expresses, in fact, a purely ideal connection between animals, the intellectual conception which unites them in the creative thought. It seems to me that the more we examine the true significance of this kind of group, the more we shall be convinced that they are not founded upon material relations. The lesser divisions which succeed next are founded upon special qualifications of the plan and differ one from the other by the character of these qualifications. Should it be found that the features in the animal kingdom which, next to the plan of structure, extend over the largest divisions are those which determine their rank or respective standing, it would appear natural to consider the orders as the second most important category in the organization of animals. Experience, however, shows that this is not the case; that the manner in which the plan of structure is executed leads to the distinction of more extensive divisions (the classes) than those which are based upon the complication of structure (the orders). As a classification can be natural only as far as it expresses real relations observed in nature, it follows therefore that classes take the second position in a system immediately under the branches. We shall see below that orders follow next, as they constitute naturally groups that are more comprehensive than families, and that we are not at liberty to invert their respective position, nor to transfer the name of one of these divisions to the other at our own pleasure, as so many naturalists are constantly doing.[7]

SECTION II

CLASSES OF ANIMALS

Before Cuvier had shown that the whole animal kingdom is constructed upon four different plans of structure, classes were the highest groups acknowledged in the systems of Zoology, and naturalists very early understood upon what this kind of division should be founded, in order to be natural, even though in practice they did not always perceive the true value of the characters upon which they established their standard of relationship. Linnæus, the first expounder of the system of animals, already distinguishes by anatomical characters the classes he has adopted, though very imperfectly; and ever since, systematic writers have aimed at drawing a more and more complete picture of the classes of animals, based upon a more or less extensive investigation of their structure.

Structure, then, is the watchword for the recognition of classes,

[7] [In the preceding discussion Agassiz's insistence on the ultimate rationality of nature as demonstrated by an underlying "plan" is exemplified. This traditional and idealistic viewpoint held that species and genera enjoyed a "real" existence in that they symbolized that fundamental reality, the categories and forms of divine thought. The essential task of the naturalist as empiricist was to describe and analyze the individual representations in nature of such higher forms.]

and an accurate knowledge of their anatomy the surest way to discover their natural limits. And yet, with this standard before them naturalists have differed and differ still greatly in the limits they assign to classes and in the number of them they adopt. It is really strange that, applying apparently the same standard to the same objects, the results of their estimation should so greatly vary; and it was this fact which led me to look more closely into the matter and to inquire whether, after all, the seeming unity of standard was not more a fancied than a real one. Structure may be considered from many points of view: first, with reference to the plan adopted in framing it; secondly, with reference to the work to be done by it, and to the ways and means employed in building it up; thirdly, with reference to the degrees of perfection or complication it exhibits, which may differ greatly, even though the plan be the same, and the ways and means employed in carrying out such a plan should not differ in the least; fourthly, with reference to the form of the whole structure and its parts, which bears no necessary relation, at all events no very close relation, to the degree of perfection of the structure, nor to the manner in which its plan is executed, nor to the plan itself, as a comparison between Bats and Birds, between Whales and Fishes, or between Holothurians and Worms may easily show; fifthly and lastly, with reference to its last finish, to the execution of the details in the individual parts.

It would not be difficult to show that the differences which exist among naturalists in their limitation of classes have arisen from an indiscriminate consideration of the structure of animals, in all these different points of view, and an equally indiscriminate application of the results obtained, to characterizing classes. Those who have not made a proper distinction between the plan of a structure and the manner in which that plan is actually executed have either overlooked the importance of the great fundamental divisions of the animal kingdom, or they have unduly multiplied the number of these primary divisions, basing their distinctions upon purely anatomical considerations, that is to say, not upon differences in the character of the general plan of structure, but upon the material development of that plan. Those, again, who have confounded the complication of the structure with the ways and means by which life is maintained through any given combination of systems of organs have failed in establishing a proper difference between class and ordinal characters, and have again and again raised orders to the rank of classes. For we shall see presently that natural orders must be based upon the different degrees of complication of structure, exhibited within the limits of the classes, while the classes themselves are characterized by the manner in which the plan of the type is carried out, that is to say, by the various combinations of the systems of organs constituting the body of the representatives of any of the great types of the animal kingdom; or perhaps, still more distinctly, the classes are characterized by the different ways in which life is maintained, and the different means employed in establishing these ways. An example will suffice to show that this distinction implies a marked difference between class and ordinal characters.

Let us compare the Polyps and Acalephs as two classes, without allowing ourselves to be troubled by the different limits assigned to them by different authors. Both are constructed upon the same plan and belong on that account to the type of Radiata. In establishing this fact we do not consider the actual structure of these animals, whether they have a nervous system or not, whether they have organs of senses or not, whether their muscles are striated or smooth, whether they have a solid frame or an entirely soft body, whether their alimentary cavity has only one opening or two opposite openings, whether it has glandular annexes or not, whether the digested food is distributed in the body one way or another, whether the undigested materials are rejected through the mouth or not, whether the sexes are distinct or not, whether they reproduce themselves only by eggs, or by budding also, whether they are simple or not: all we need know, in order to refer them to the branch of Radiata, is whether the plan of their structure exhibits a general radiated arrangement or not. But when we would distinguish Polyps, Acalephs, and Echinoderms as classes, or rather, when we would ascertain what are the classes among Radiata, and how many there are, we must inquire into the manner in which this idea of radiation, which lies at the foundation of their plan of structure, is actually expressed in all the animals exhibiting such a plan; and we find easily that while in some (the Polypi) the body exhibits a large cavity, divided by radiating partitions into a number of chambers, into which hangs a sac (the digestive cavity), open below, so as to pour freely the digested food into the main cavity, whence it is circulated to and fro in all the chambers by the agency of vibrating cilia; in others (the Acalephs) the body is plain and full, not to be compared to a hollow sac, traversed only in its thickness by radiating tubes which arise from a central cavity (the digestive cavity) without a free communication with one another for their whole length, etc., etc.; while in others still (the Echinoderms) there is a tough or rigid envelope to the body, inclosing a large cavity in which are contained a variety of distinct systems of organs, etc.

Without giving here a full description of these classes, I only wish to show that what truly characterizes them is not the complication of their structure (for Hydroid Medusæ are hardly more complicated in their structure than Polyps), but the manner in which the plan of Radiata is carried out, the ways in which life is maintained in these animals, the means applied to this end; in one word, the combinations of their structural elements. But the moment we would discern what are the orders of these classes these considerations no longer suffice; their structure has to be viewed in a different light; it is now the complication of these apparatus which may guide us. Actinarians and Halcyonarians among Polypi, as orders, differ, the first by having a larger and usually indefinite number of simple tentacles, an equally large number of internal partitions, etc., while in Halcyonarians the eight tentacles are lobed and complicated, and all the parts are combined in pairs, in definite numbers, etc., differences which establish a distinct standing between them in their class, assigning the latter a higher rank than the former.

It follows, then, from the preceding remarks that classes are to be distinguished by the manner in which the plan of their type is executed, by the ways and means by which this is done, or, in other words, by the combinations of their structural elements, that is to say, by the combinations of the different systems of organs building up the body of their representatives. We need not consider here the various forms under which the structure is embodied, nor the ultimate details, nor the last finish which this structure may exhibit, as a moment's reflection will convince any one that neither form nor structural details can ever be characteristic of classes.

There is another point to which I would call attention respecting the characteristics of classes. These great divisions, so important in the study of the animal kingdom that a knowledge of their essential features is rightly considered as the primary object of all investigations in comparative anatomy, are generally represented as exhibiting each some essential modification of the type to which they belong. This view, again, I consider to be a mistaken appreciation of the facts to which Cuvier has already called attention, though his warning has remained unnoticed.[8] There is in reality no difference in the plan of animals belonging to different classes of the same branch. The plan of structure of Polypi is no more a modification of that of Acalephæ, than that of Acalephæ or Echinoderms is a modification of the plan of Polyps; the plan is exactly the same in all three; it may be represented by one simple diagram and may be expressed in one single word, radiation; it is the manifestation of one distinct, characteristic idea. But this idea is exhibited in nature under the most different forms and expressed in different ways by the most diversified combinations of structural modifications and in the most varied relations. In the innumerable representatives of each branch of the animal kingdom it is not the plan that differs, but the manner in which this plan is executed. In the same manner as the variations played by a skilful artist upon the simplest tune are not modifications of the tune itself, but only different expressions of the same fundamental harmony, just so are neither the classes, nor the orders, nor the families, nor the genera, nor the species of any great type, modifications of its plan, but only its different expressions, the different ways in which the fundamental thought embodied in it is manifested in a variety of living beings.

In studying the characteristics of classes we have to deal with structural features, while in investigating their relations to the branches of the animal kingdom to which they belong we have only to consider the general plan, the framework, as it were, of that structure, not the structure itself. This distinction leads to an important practical result. Since in the beginning of this century naturalists have begun, under the lead of the German physiophilosophers,[9] to

[8] *Règne animal* (2d ed.), I, 48.

[9] [Agassiz is commenting on the early nineteenth-century views of organic creation identified primarily with German speculative biology and also known as *Naturphilosophie*. Goethe and Oken were leading advocates of this idealized view of nature, and Agassiz was very much impressed by it while a student in Germany. In later life he was critical of this viewpoint as tending to support a concept of development from lower to higher forms in a unified fashion. This ambivalence can be seen in his subsequent analysis of Oken's system of classification.]

compare more closely the structure of the different classes of the animal kingdom, points of resemblance have been noticed between them which had entirely escaped the attention of earlier investigators, structural modifications have been identified, which at first seemed to exhibit no similarity, so much so, that step by step these comparisons have been extended over the whole animal kingdom, and it has been asserted that whatever may be the apparent differences in the organization of animals, they should be considered as constructed of parts essentially identical. This assumed identity of structure has been called homology.[10] But the progress of science is gradually restricting these comparisons within narrower limits, and it appears now that the structure of animals is homologous only as far as they belong to the same branch, so much so, that the study of homologies is likely to afford one of the most trustworthy means of testing the natural limits of any of the great types of the animal kingdom. While, however, homologies show the close similarity of apparently different structures and the perfect identity of their plan within the same branches of the animal kingdom, yet they daily exhibit more and more striking differences, both in plan and structure, between the branches themselves, leading to the suspicion that systems of organs which are generally considered as identical in different types will, in the end, prove essentially different, as, for instance, the so-called gills in Fishes, Crustacea, and Mollusks.

It requires no great penetration to see already that the gills of Crustacea are homologous with the tracheæ of Insects and the so-called lungs of certain spiders, in the same manner as the gills of aquatic Mollusks are homologous with the so-called lungs of our air-breathing snails and slugs. Now, until it can be shown that all these different respiratory organs are truly homologous, I hold it to be more natural to consider the system of respiratory organs in Mollusks, in Articulates, and in Vertebrates as essentially different among themselves, though homologous within the limits of each type; and this remark I would extend to all their systems of organs, to their solid frame, to their nervous system, to their muscular system, to their digestive apparatus, to their circulation, and to their reproductive organs, etc. It would not be difficult to show now that the alimentary canal with its glandular appendages in Vertebrata is formed

[10] See Chap. I, Sect. v.

in an entirely different way from that of Articulates or Mollusks, and that it cannot be considered as homologous in all these types. And if this be true, we must expect soon an entire reform of our methods of illustrating comparative anatomy.

Finally, it ought to be remembered, in connection with the study of classes as well as that of other groups, that the amount of difference existing between any two divisions is nowhere the same. Some features in nature seem to be insisted upon with more tenacity than others, to be repeated more frequently and more widely, and to be impressed upon a larger number of representatives. This unequal weight of different groups, so evident everywhere in the animal kingdom, ought to make us more cautious in estimating their natural limits and prevent us from assigning an undue value to the differences observed between living beings, never overrating apparently great discrepancies nor underrating seemingly trifling variations. The right path, however, can only be ascertained by extensive investigations made with special reference to this point.

Everybody must know that the males and females of some species differ much more one from the other than many species do, and yet the amount of difference observed between species is constantly urged, even without a preliminary investigation, as an argument for distinguishing them. These differences, moreover, are not only quantitative, they are to a still greater extent also qualitative. In the same manner do genera differ more or less one from the other, even in the same family; and such inequality, and not an equable apportionment, is the norm throughout nature. In classes it is not only exhibited in the variety of their forms, but also to an extraordinary extent in their numbers, as, for instance, in the class of Insects compared to that of Worms or Crustacea. The primary divisions of the animal kingdom differ in the same manner one from the other. Articulata are by far the most numerous branch of the whole animal kingdom; their number exceeding greatly that of all other animals put together. Such facts are in themselves sufficient to show how artificial classifications must be which admit only the same number and the same kind of divisions for all the types of the animal kingdom.

SECTION III

ORDERS AMONG ANIMALS

Great as is the discrepancy between naturalists respecting the number and limits of classes in the animal kingdom, their disagreement in regard to orders and families is yet far greater. These conflicting views, however, do not in the least shake my confidence in the existence of fixed relations between animals, determined by thoughtful considerations. I would as soon cease to believe in the existence of one God because men worship Him in so many different ways or because they even worship gods of their own making, as to distrust the evidence of my own senses respecting the existence of a pre-established and duly considered system in nature, the arrangement of which preceded the creation of all things that exist.

From the manner in which orders are generally characterized and introduced into our systems it would seem as if this kind of group were interchangeable with families. Most botanists make no difference even between orders and families and take almost universally the terms as mere synonyms. Zoologists have more extensively admitted a difference between them, but while some consider the orders as superior, others place families higher; others admit orders without at the same time distinguishing families and *vice versa* introduce families into their classification without admitting orders; others still admit tribes as intermediate groups between orders and families. A glance at any general work on Zoology or Botany may satisfy the student how utterly arbitrary the systems are in this respect. The *Règne animal* of Cuvier exhibits even the unaccountable feature that while orders and families are introduced in some classes,[11] only orders are noticed in others,[12] and even some exhibit only a succession of genera under the head of their class, without any further

[11] In the classes Mammalia, Birds, Reptiles, and Fishes Cuvier distinguishes mostly families as well as orders. In the class of Mammalia some orders number no families, whilst others are divided into tribes instead of families. In the class of Gasteropods, Annelids, Intestinal Worms, and Polyps, some of the orders only are divided into families, while the larger number are not.

[12] The classes Echinoderms, Acalephs, and Infusoria are divided into orders, but without families.

grouping among them into orders or families.[13] Other classifications exhibit the most pedantic uniformity of a regular succession in each class, of sub-classes, orders, sub-orders, families, sub-families, tribes, sub-tribes, genera, sub-genera, divisions, sections, and sub-divisions, sub-sections, etc., but bear upon their face that they are made to suit preconceived ideas of regularity and symmetry in the system, and that they are by no means studied from nature.

To find out the natural characters of orders from that which really exists in nature I have considered attentively the different systems of Zoology in which orders are admitted and apparently considered with more care than elsewhere, and in particular the *Systema Naturæ*[14] of Linnæus, who first introduced in Zoology that kind of group, and the works of Cuvier, in which orders are frequently characterized with unusual precision; and it has appeared to me that the leading idea prevailing everywhere respecting orders, where these groups are not admitted at random, is that of a definite rank among them, the desire to determine the relative standing of these divisions, to ascertain their relative superiority or inferiority, as the name order adopted to designate them already implies. The first order in the first class of the animal kingdom, according to the classification of Linnæus, is called by him *Primates,* expressing, no doubt, his conviction that these beings, among which Man is included, rank uppermost in their class. Blainville uses here and there the expression of "degrees of organization," to designate orders. It is true Lamarck uses the same expression to designate classes. We find therefore here, as everywhere, the same vagueness in the definition of the different kinds of groups adopted in our systems. But if we would give up any arbitrary use of these terms and assign to them a definite scientific meaning, it seems to me most natural and in accordance with the practice of the most successful investigators of the animal kingdom, to call orders such divisions as are characterized by different degrees of complication of their structure, within the limits of the classes. As such I would consider, for instance, the Actinoids and Halcyonoids in the class of Polypi, as circumscribed by Dana; the Hydroids, the Disco-

[13] Such are his classes of Cephalopods, Pteropods, Brachiopods, and Cirripeds (Cirrhopods). Of the Cephalopods, he says, however, they constitute but one order, and he calls them a family, and yet he distinguishes them as a class. *Règne animal* (2d ed.), III, 8, 11, 22.

[14] [Leiden (1735); 12th ed., 3 vols., Stockholm (1766–1768).]

phoræ, and the Ctenoids among Acalephs; the Crinoids, Asterioids, Echinoids, and Holothuriæ among Echinoderms; the Bryozoa, Brachiopods, Tunicata, Lamellibranchiata among Acephala; the Branchifera and Pulmonata among Gasteropods; the Ophidians, the Saurians, and the Chelonians among Reptiles; the Ichthyoids and the Anoura among Amphibians, etc.

Having shown in the preceding paragraph that classes rank next to branches, it would be proper I should show here that orders are natural groups which stand above families in their respective classes; but for obvious reasons I have deferred this discussion to the following paragraph, which relates to families, as it will be easier for me to show what is the respective relation of these two kinds of groups after their special character has been duly considered.

From the preceding remarks respecting orders it might be inferred that I deny all gradation among all other groups, or that I assume that orders constitute necessarily one simple series in each class. Far from asserting any such thing, I hold on the contrary, that neither is necessarily the case. But to explain fully my views upon this point I must introduce here some other considerations. It will be obvious, from what has already been said (and the further illustration of this subject will only go to show to what extent this is true) that there exists an unquestionable hierarchy between the different kinds of groups admitted in our systems, based upon the different kinds of relationship observed among animals; that branches are the most comprehensive divisions, including each several classes; that orders are subdivisions of the classes, families subdivisions of orders, genera subdivisions of families, and species subdivisions of the genera; but not in the sense that each type should necessarily include the same number of classes, nor even necessarily several classes, as this must depend upon the manner in which the type is carried out. A class, again, might contain no orders,[15] if its representatives presented no different degrees characterized by the greater or less complication of their structure; or it may contain many or few, as these gradations are more or less numerous and well marked; but as the representatives of any and every class have of necessity a definite form, each class must contain at least one family or many families, indeed, as many as there are systems of forms under which its representatives may be

[15] See Chap. I, Sect. 1.

combined, if form can be shown to be characteristic of families. The same is the case with genera and species; and nothing is more remote from the truth than the idea that a genus is better defined in proportion as it contains a greater number of species, or that it may be necessary to know several species of a genus before its existence can be fully ascertained. A genus may be more satisfactorily characterized, its peculiarity more fully ascertained, its limits better defined, when we know all its representatives; but I am satisfied that any natural genus may be at least pointed out, however numerous its species may be, from the examination of any single one of them. Moreover, the number of genera, both in the animal and vegetable kingdom, which contain but a single species is so great that it is a matter of necessity in all these cases to ascertain their generic characteristics from that one species. Again, such species require to be characterized with as much precision, and their specific characters to be described with as much minuteness, as if a host of them, but not yet known, existed besides. It is a very objectionable practice among zoologists and botanists to remain satisfied in such cases with characterizing the genus, and perhaps to believe what some writers have actually stated distinctly, that in such cases generic and specific characters are identical.

Such being the natural relations and the subordination of types, classes, orders, families, genera, and species, I believe, nevertheless, that neither types, nor classes (orders of course not at all) nor families, nor genera, nor species have the same standing when compared among themselves. But this does not in the least interfere with the prominent features of orders, for the relative standing of types, or classes, or families, or genera, or species does not depend upon the degrees of complication of their structures as that of orders does, but upon other features, as I will now show. The four great types or branches of the animal kingdom, characterized as they are by four different plans of structure, will each stand higher or lower, as the plan itself bears a higher or lower character, and that this may be the case we need only compare Vertebrata and Radiata.[16] The different classes of one type will stand higher or lower, as the ways in which and the means with which the plan of the type to which they

[16] I must leave out the details of such comparisons, as a mere mention of the point suffices to suggest them; moreover, any text-book of comparative anatomy may furnish the complete evidence to that effect.

belong is carried out are of a higher or lower nature. Orders in any or all classes are of course higher or lower according to the degree of perfection of their representatives, or according to the complication or simplicity of their structure. Families may stand higher or lower as the peculiarities of their form are determined by modifications of more or less important systems of organs. Genera may stand higher or lower as the structural peculiarities of the parts constituting the generic characteristics exhibit a higher or lower grade of development. Species, lastly, may stand one above the other, in the same genus, according to the character of their relations to the surrounding world, or that of their representatives to one another. These remarks must make it plain that the respective rank of groups of the same kind among themselves must be determined by the superior or inferior grade of those features upon which they are themselves founded; while orders alone are strictly defined by the natural degrees of structural complications exhibited within the limits of the classes.

As to the question whether orders constitute necessarily one simple series in their respective classes, I would say that this must depend upon the character of the class itself or the manner in which the plan of the type is carried out within the limits of the class. If the class is homogeneous, that is, if it is not primarily subdivided into subclasses, the orders will of course form a single series; but if some of its organic systems are developed in a different way from the others, there may be one or several parallel series, each subdivided into graduated orders. This can of course only be determined by a much more minute study of the characteristics of classes than has been made thus far, and mere guesses at such an internal arrangement of the classes into series, as those proposed by Kaup or Fitzinger,[17] can only be considered as the first attempts towards an estimation of the relative value of the intermediate divisions which may exist between the classes and their orders.

Oken and the physiophilosophers generally have taken a different view of orders. Their idea is that orders represent in their respective classes the characteristic features of the other types of the animal kingdom. As Oken's Intestinal or Gelatinous animals are character-

[17] [Johann J. Kaup, *Classification der Säugethiere und Vögel* (Darmstadt, 1844); Leopold Fitzinger, *Systema Reptilium* (Vienna, 1843).]

ized by a single system of organs, the intestine, they contain no distinct orders, but each class has three tribes, corresponding to the three classes of this type, which are Infusoria, Polypi, and Acalephs. The tribes of the class of Infusoria are Infusoria proper, Polypoid Infusoria, and Acalephoid Infusoria; the tribes of the class of Polypi are Infusorial Polypi, Polypi proper, and Acalephoid Polypi; the tribes of the class Acalephs are Infusorial Acalephs, Polypoid Acalephs, and Acalephs proper. But the classes of Mollusks which are said to be characterized by two systems of organs, the intestine and the vascular system, contain each two orders, one corresponding to the Intestinal animals, the other to the type of Mollusks, and so Acephala are divided into the order of Gelatinous Acephala and that of Molluscoid Acephala, and the Gasteropods and Cephalopods in the same manner into two orders each. The Articulata are considered as representing three systems of organs, the intestinal, the vascular, and the respiratory systems; hence their classes are divided each into three orders. For instance, the Worms contain an order of Gelatinous Worms, one of Molluscoid Worms, one of Annulate Worms, and the same orders are adopted for Crustacea and Insects. Vertebrata are said to represent five systems, the three lower ones being the intestine, the vessels, and the respiratory organs, the two higher the flesh (that is, bones, muscles, and nerves) and the organs of senses; hence, five orders in each class of this type, as, for example, Gelatinous Fishes, Molluscoid Fishes, Entomoid Fishes, Carnal Fishes, and Sensual Fishes, and so also in the class of Reptiles, Birds, and Mammalia.[18]

I have entered into so many details upon these vagaries of the distinguished German philosopher, because these views, however crude, have undoubtedly been suggested by a feature of the animal kingdom which has thus far been too little studied: I mean the analogies which exist among animals, besides their true affinities, and which cross and blend, under modifications of strictly homological structures, other characters which are only analogical. But it seems to me that the subject of analogies is too little known, the facts bearing upon this kind of relationship being still too obscure to be taken as the basis of such important groups in the animal kingdom as the

[18] See further developments upon this subject in Oken, *Naturphilosophie,* and *Allgemeine Naturgeschichte* (14 vols., Stuttgart, 1833–1843), IV, 582.

orders are, and I would insist upon considering the complication or gradation of structure as the feature which should regulate their limitation, if under order we are to understand natural groups expressing the rank, the relative standing, the superiority or inferiority of animals in their respective classes. Of course groups thus characterized cannot be considered as mere modifications of the classes, being founded upon a special category of features.

SECTION IV

FAMILIES

Nothing is more indefinite than the idea of form, as applied by systematic writers, in characterizing animals. Here, it means a system of the most different figures having a common character, as, for instance, when it is said of Zoophytes that they have a radiated form; there, it indicates any outline which circumscribes the body of animals, when, for instance, animal forms are alluded to in general, instead of designating them simply as animals; here, again, it means the special figure of some individual species. There is in fact no group of the animal kingdom, however extensive or however limited, from the branches down to the species, in which the form is not occasionally alluded to as characteristic. Speaking of Articulates, C. E. v. Baer characterizes them as the type with elongated forms; Mollusks are to him the type with massive forms; Radiates that with peripheric symmetry; Vertebrates that with double symmetry, evidently taking their form in its widest sense as expressing the most general relations of the different dimensions of the body to one another. Cuvier speaks of form in general with reference to these four great types as a sort of mould, as it were, in which the different types would seem to have been cast. Again, form is alluded to in characterizing orders; for instance, in the distinction between the Brachyourans and the Macrourans among Crustacea, or between the Saurians, the Ophidians, and the Chelonians. It is mentioned as a distinguishing feature in many families, e.g. the Cetacea, the Bats, etc. Some genera are separated from others in the same family on the ground of differences of form; and in almost every description of species, especially when they are considered isolatedly, the form is described at full length. Is there not in this indiscriminate use of the term of form a confusion of ideas, a want of precision in the estimation of what ought to be called form and what might be designated by another name? It seems to me to be the case. In the first place, when form is considered as characteristic of Radiata or Articulata or any other of the great types of the animal kingdom, it is evident that it is not a definite outline and well-determined figure which is meant, but that here the word form is used as synonym for plan. Who, for instance, would describe the tubular body of an Holothuria as characterized by a form similar to that of the Euryale, or that of an Echinus as identical with that of an Asterias? And who does not see that, as far as the form is concerned, Holothuriæ resemble Worms much more than they resemble any other Echinoderm, though as far as the plan of their structure is concerned they are genuine Radiates and have nothing to do with the Articulates?

Again, a superficial glance at any and all the classes of the animal kingdom is sufficient to show that each contains animals of the most diversified forms. What can be more different than Bats and Whales, Herons and Parrots, Frogs and Sirens, Eels and Turbots, Butterflies and Bugs, Lobsters and Barnacles, Nautilus and Cuttlefishes, Slugs and Conchs, Clams and compound Asidians, Pentacrinus and Spatangus, Beroe and Physalia, Actinia and Gorgonia? And yet they belong respectively to the same class, as they are coupled here: Bats and Whales together, etc. It must be obvious then that form cannot be a characteristic element of classes, if we would understand anything definite under that name.

But form has a definite meaning understood everywhere when applied to well known animals. We speak, for instance, of the human form; an allusion to the form of a horse or that of a bull conveys at once a distinct idea; everybody would acknowledge the similarity of form of the horse and ass and knows how to distinguish them by their form from dogs or cats, or from seals and porpoises. In this definite meaning form corresponds also to what we call figure when speaking of men and women, and it is when taken in this sense that I would now consider the value of forms as characteristic of different animals. We have seen that form cannot be considered as a character of branches nor of classes; let us now examine further whether it is a character of species. A rapid review of some of the best known types of the animal kingdom, embracing well-defined genera with many species, will at once show that this cannot be the case, for such species do not generally show the least difference in their forms. Neither the many species of Squirrels, nor the true Mice, nor the Weasels, nor the Bears, nor the Eagles, nor the Falcons, nor the Sparrows, nor the Warblers, nor the genuine Woodpeckers, nor the true Lizards, nor the Frogs, nor the Toads, nor the Skates, nor the Sharks proper, nor the Turbots, nor the Soles, nor the Eels, nor the Mackerels, nor the Sculpins, nor the genuine Shrimps, nor the Crawfishes, nor the Hawk-moths, nor the Geometers, nor the Dorbugs, nor the Spring-Beetles, nor the Tapeworms, nor the Cuttlefishes, nor the Slugs, nor the true Asterias, nor the Sea-Anemones could be distinguished among themselves, one from the other, by their form only. There may be differences in the proportions of some of their parts, but the pattern of every species belonging to well-defined natural genera is so completely identical that it will never afford specific characters. There are genera in our system which, as they now stand, might be alluded to as examples contrary to this statement; but such genera are still based upon very questionable features and are likely to be found in the end to consist of unnatural associations of heterogeneous species: at all events, all recent improvements in Zoology have gone to limit genera gradually more and more in such a manner that the species belonging to each have shown successively less and less difference in form, until they have assumed in that respect the most homogeneous appearance. Are natural genera any more to be distinguished by their form one from the other? Is there any appreciable difference in the general form? I say purposely general form, because a more or less prominent nose, larger or smaller ears, longer or shorter claws, etc., do not essentially modify the form. Is there any real difference in the general form between the genera of the most natural families? Do, for instance, the genera of Ursina, the Bears, the Badger, the Wolverines, the Raccoons, differ in form? Do the Phocoidæ, the Delphinoidæ, the Falconinæ, the Turdinæ, the Fringillinæ, the Picinæ, the Scolopacinæ, the Chelonioidæ, the Geckonina, the Colubrina, the Sparoidæ, the Elateridæ, the Pyralidoidæ, the Echinoidæ, etc., differ any more among themselves? Certainly not;

though to some extent there are differences in the form of the representatives of one genus when compared to those of another genus; but when rightly considered, these differences appear only as modifications of the same type of forms. Just as there are more or less elongated ellipses, so do we find the figure of the Badgers somewhat more contracted than that of either the Bears, or the Raccoons, or the Wolverines, that of the Wolverines somewhat more elongated than that of the Raccoons; but the form is here as completely typical as it is among the Viverrina, or among the Canina, or among the Bradypodidæ, or among the Delphinoidæ, etc., etc. We must, therefore, exclude form from the characteristics of natural genera, or at least introduce it only as a modification of the typical form of natural families.

Of all the natural groups in the animal kingdom there remain then only families and orders, for the distinction of which form can apply as an essential criterion. But these two kinds of groups are just those upon which zoologists are least agreed, so that it may not be easy to find a division which all naturalists would agree to take as an example of a natural order. Let us, however, do our best to settle the difficulty and suppose for a moment that what has been said above respecting the orders is well founded, that orders are natural groups characterized by the degree of complication of their structure and expressing the respective rank of these groups in their class, then we shall find less difficulty in pointing out some few groups which could be generally considered as orders. I suppose most naturalists would agree, for instance, that among Reptiles the Chelonians constitute a natural order; that among Fishes, Sharks and Skates constitute an order also; and if any one would urge the necessity of associating also the Cyclostomes with them, it would only the better serve my purposes. Ganoids, even circumscribed within narrower limits than those I had assigned to them, and perhaps reduced to the extreme limits proposed for them by J. Müller, I am equally prepared to take as an example, though I have in reality still some objections to this limitation, which, however, do not interfere with my present object. Decapods, among Crustacea, I suppose everybody would also admit as an order, and I do not care here what other families are claimed besides Decapods to complete the highest order of Crustacea. Among Acephala, I trust Bryozoa, Tunicata, Brachiopods, and Lamellibranchiata would be also very generally considered to be natural orders. Among Echinoderms, I suppose Crinoids, Asterioids, Echinoids, and Holothurioids would be conceded also as such natural orders; among Acalephs the Beroids, and perhaps also Discophoræ and Hydroids; while among the Polypi, the Halcyonoids constitute a very natural order when compared with the Actinoids.

Let us now consider these orders with reference to the characteristic forms they include. The forms of the genuine Testudo, of Trionyx, and of Chelonia are very different, one from the other, and yet few orders are so well circumscribed as that of Chelonians. The whole class of Fishes scarcely exhibits greater differences than those observed in the forms of the common Sharks, the Sawfishes, the common Skates, and the Torpedo, not to speak of the Cyclostomes and Myxinoids, if these families were also considered as members of the order of Placoids. Ganoids cannot be circumscribed within narrower limits than those assigned to them by J. Müller, and yet this order, thus limited, contains forms as heterogeneous as the Sturgeons, the Lepidosteus, the Polypterus, the Amia, and a host of extinct genera and families, not to speak of those families I had associated with them and which Prof. Müller would have removed, which, if included among Ganoids, would add still more heteromorphous elements to this order. Among Decapods we need only remember the Lobsters and Crabs to be convinced that it is not similarity of form which holds them so closely together as a natural order. How heterogeneous Bryozoa, Brachiopods, and Tunicata are among themselves, as far as their form is concerned, everybody knows who has paid the least attention to these animals.

Unless, then, form be too vague an element to characterize any kind of natural group in the animal kingdom, it must constitute a prominent feature of families. I have already remarked that orders and families are the groups upon which zoologists are least agreed and to the study and characterizing of which they have paid least attention. Does this not arise simply from the fact that, on the one hand, the difference between ordinal and class characters has not been understood and only assumed to be a difference of degree; and, on the other hand, that the importance of the form, as the prominent character of families, has been entirely overlooked? For, though so few natural families of animals are well characterized, or characterized at all, we cannot open a modern treatise upon any class of animals without finding the genera more or less naturally grouped together, under the heading of a generic name with a termination in *idæ* or *inæ* indicating family and sub-family distinctions; and most of these groups, however unequal in absolute value, are really natural groups, though far from designating always natural families, being as often orders or sub-orders as families or sub-families. Yet they indicate the facility there is, almost without study, to point out the intermediate natural groups between the classes and the genera. This arises in my opinion from the fact that family resemblance in the animal kingdom is most strikingly expressed in the general form, and that form is an element which falls most easily under our perception, even when the observation is made superficially. But at the same time, form is most difficult to describe accurately, and hence the imperfection of most of our family characteristics and the constant substitution for such characters of features which are not essential to the family. To prove the correctness of this view I would only appeal to the experience of every naturalist. When we see new animals, does not the first glance, that is, the first impression made upon us by their form, give us at once a very correct idea of their nearest relationship? We perceive, before examining any structural character, whether a Beetle is a Carabicine, a Longicorn, an Elaterid, a Curculionid, a Chrysomeline; whether a Moth is a Noctuelite, a Geometrid, a Pyralid, etc.; whether a bird is a Dove, a Swallow, a Humming-bird, a Woodpecker, a Snipe, a Heron, etc., etc. But before we can ascertain its genus we have to study the structure of some characteristic parts; before we can combine families into natural groups we have to make a thorough investigation of their whole structure and compare it with that of other families. So form is characteristic of families; and I can add, from a careful investigation of the subject for several years past, during which I have reviewed the whole animal kingdom with reference to this and other topics connected with classification, that form is the essential characteristic of families.[19] I do not mean the mere outline, but form as determined

[19] These investigations, which have led to most interesting results, have delayed thus far the publication of the systematic part of the *Principles of Zoology* (Pt. I, 1848), undertaken in common with my friend, Dr. Augustus A. Gould, and which I would not allow to appear before I could revise the whole animal kingdom in this new light,

by structure; that is to say, that families cannot be well defined nor circumscribed within their natural limits without a thorough investigation of all those features of the internal structure which combine to determine the form.

The characteristic of the North American Chelonians[20] may serve as an example how this subject is to be treated. I will only add here that, however easy it is at first, from the general impression made upon us by the form of animals, to obtain a glimpse of what may fairly be called families, few investigations require more patient comparisons than those by which we ascertain the natural range of modifications of any typical form and the structural features upon which it is based. Comparative anatomy has so completely discarded every thing that relates to Morphology; the investigations of anatomists lean so uniformly towards a general appreciation of the connections and homologies of the organic systems which go to build up the body of animals, that for the purpose of understanding the value of forms and their true foundation they hardly [if] ever afford any information, unless it be here and there a consideration respecting teleological relations.

Taking for granted that orders are natural groups characterized by the complication of their structure and that the different orders of a class express the different degrees of that complication; taking now further for granted that families are natural groups characterized by their form as determined by structural peculiarities, it follows that orders are the superior kind of division, as we have seen that the several natural divisions which are generally considered as orders contain each several natural groups, characterized by different forms, that is to say, constituting as many distinct families.

After this discussion it is hardly necessary to add that families cannot by any means be considered as modifications of the orders to which they belong, if orders are to be characterized by the degrees of complication of their structure and families by their forms. I would also further remark that there is one question relating to the form of animals which I have not touched here and which it is still more important to consider in the study of plants, namely, the mode

---

in order to introduce as much precision as possible in its classification. [This work was never completed.]

[20] See my *Contributions to the Natural History of the United States,* I, 317–366.

of association of individuals into larger or smaller communities as we observe them, particularly among Polyps and Acalephs. These aggregations have not, as far as their form is concerned, the same importance as the form of the individual animals of which they are composed, and therefore seldom afford trustworthy family characters. But this point may be more appropriately considered in connection with the special illustration of our Hydroids, to which my next volume[21] is to be devoted.

I have stated above that botanists have defined the natural families of plants with greater precision than zoologists those of animals; I have further remarked also that most of them make no distinction between orders and families. This may be the result of the peculiar character of the vegetable kingdom, which is not built upon such entirely different plans of structure as are animals of different branches. On the contrary, it is possible to trace among plants a certain gradation between their higher and lower types more distinctly than among animals, even though they do not, any more than animals, constitute a simple series. It seems to me, nevertheless, that if Cryptogams, Gymnosperms, Monocotyledons, and Dicotyledons can be considered as branches of the vegetable kingdom, analogous to Radiata, Mollusks, Articulata, and Vertebrata among animals, such divisions as Fungi, Algæ, Lichens, Mosses, Hepaticæ, and Ferns in the widest sense may be taken as classes. Diatomaceæ, Confervæ, and Fuci may then be considered as orders; Mosses and Hepaticæ as orders; Equisetaceæ, Ferns proper, Hydropterids, and Lycopodiaceæ as orders also; as they exhibit different degrees of complication of structure while their natural subdivisions, which are more closely allied in form or habitus, may be considered as families; natural families among plants having generally as distinct a port, as families among animals have a distinct form. We need only remember the Palms, the Coniferæ, the Umbelliferæ, the Compositæ, the Leguminosæ, the Labiatæ, etc., as satisfactory examples of this kind.

[21] [See *Contributions* . . . III (1860).]

SECTION V

GENERA

Linnæus already knew very well that genera exist in nature, though what he calls genera constitute frequently groups to which we give at present other names, as we consider many of them as families; but it stands proved by his writings that he had fully satisfied himself of the real existence of such groups, for he says distinctly in his *Philosophia Botanica,* sect. 169, "Scias characterem non constituere genus, sed genus characterem. Characterem fluere e genere, non genus e charactere. Characterem non esse, ut genus fiat, sed ut genus noscatur." [22]

It is surprising that notwithstanding such clear statements, which might have kept naturalists awake respecting the natural foundation of genera, such loose ideas have become prevalent upon this subject, that at present the number of investigators who exhibit much confidence in the real existence of their own generic distinctions is very limited. And as to what genera really are, the want of precision of ideas appears still greater. Those who have considered the subject at all seem to have come to the conclusion that genera are nothing but groups including a certain number of species agreeing in some more general features than those which distinguish species; thus recognizing no difference between generic and specific characters as such, as a single species may constitute a genus whenever its characters do not agree with the characters of other species, and many species may constitute a genus because their specific characters agree to a certain extent among themselves. Far from admitting such doctrines, I hope to be able to show that, however much or however little species may differ among themselves as species, yet they may constitute a natural genus, provided their respective generic characters are identical.

I have stated before that in order to ascertain upon what the different groups adopted in our systems are founded, I consulted the works

[22] ["... characters do not form a genus but the genus constitutes the characters. Characters come from the genus, not the genus from the characters. Characters are not there so that there should be a genus, but in order that the genus should be recognized." *Philosophia Botanica* (Stockholm, 1751).]

of such writers as are celebrated in the annals of science for having characterized with particular felicity any one kind of these groups, and I have mentioned Latreille as prominent among zoologists for the precision with which he has defined the genera of Crustacea and Insects, upon which he has written the most extensive work extant.[23] An anecdote which I have often heard repeated by entomologists who knew Latreille well is very characteristic as to the meaning he connected with the idea of genera. At the time he was preparing the work just mentioned he lost no opportunity of obtaining specimens, the better to ascertain from nature the generic peculiarities of these animals, and he used to apply to the entomologists for contributions to his collection. It was not show specimens he cared to obtain, any would do, for he used to say he wanted them only "to examine their parts." Have we not here a hint from a master to teach us what genera are and how they should be characterized? Is it not the special structure of some part or other which characterizes genera? Is it not the finish of the organization of the body, as worked out in the ultimate details of structure, which distinguishes one genus from another? Latreille, in expressing the want he felt with reference to the study of genera, has given us the key-note of their harmonious relations to one another. Genera are most closely allied groups of animals, differing neither in form, nor in complication of structure, but simply in the ultimate structural peculiarities of some of their parts; and this is, I believe, the best definition which can be given of genera. They are not characterized by modifications of the features of the families, for we have seen that the prominent trait of family difference is to be found in a typical form; and genera of the same family may not differ at all in form. Nor are genera merely a more comprehensive mould than the species, embracing a wide range of characteristics; for species in a natural genus should not present any structural differences, but only such as express the most special relations of their representatives to the surrounding world and to each other. Genera, in one word, are natural groups of a peculiar kind, and their special distinction rests upon the ultimate details of their structure.

[23] *Genera Crustaceorum et Insectorum* (4 vols., Paris and Strasburg, 1806–1809).

SECTION VI

SPECIES

It is generally believed that nothing is easier than to determine species and that, of all the degrees of relationship which animals exhibit, that which constitutes specific identity is the most clearly defined. An unfailing criterion of specific identity is even supposed to exist in the sexual connection which so naturally brings together the individuals of the same species in the function of reproduction. But I hold that this is a complete fallacy, or at least a *petitio principii,* not admissible in a philosophical discussion of what truly constitutes the characteristics of species. I am even satisfied that some of the most perplexing problems involved in the consideration of the natural limits of species would have been solved long ago, had it not been so generally urged that the ability and natural disposition of individuals to connect themselves in fertile sexual intercourse was of itself sufficient evidence of their specific identity. Without alluding to the fact that every new case of hybridity[24] is an ever-returning protest against such an assertion, and without entering here into a discussion respecting the possibility or practicability of setting aside this difficulty by introducing the consideration of the limited fertility of the progeny of individuals of different species, I will only remark that as long as it is not proved that all the varieties of dogs and of any others of our domesticated animals, and of our cultivated plants, are respectively derived from one unmixed species, and as long as doubts can be entertained respecting the common origin of all races of men from one common stock, it is not logical to admit that sexual connection resulting even in fertile offspring is a trustworthy evidence of specific identity.

To justify this assertion, I would only ask, where is the unprejudiced naturalist who in our days would dare to maintain: 1st, that it is proved that all the domesticated varieties of sheep, of goats, of bulls, of llamas, of horses, of dogs, of fowls, etc., are respectively

[24] Braun, *Betrachtungen über die Erscheinung der Verjüngung;* Samuel George Morton, "Hybridity in Animals . . . ," *American Journal of Science,* III (2d ser., 1847), 39–50, 203–212, and "Additional Observations on Hybridity in Animals . . . ," *Charleston Medical Journal,* V (1850), 755–805.

derived from one common stock; 2d, that the supposition that these varieties have originated from the complete amalgamation of several primitively distinct species is out of the question; and 3d, that varieties imported from distant countries and not before brought together, such as the Shanghae fowl, for instance, do not completely mingle? Where is the physiologist who can conscientiously affirm that the limits of the fertility between distinct species are ascertained with sufficient accuracy to make it a test of specific identity? And who can say that the distinctive characters of fertile hybrids and of unmixed breeds are sufficiently obvious to enable anybody to point out the primitive features of all our domesticated animals or of all our cultivated plants? As long as this cannot be done, as long as the common origin of all races of men and of the different animals and plants mentioned above is not proved, while their fertility with one another is a fact which has been daily demonstrated for thousands of years; as long as large numbers of animals are hermaphrodites, never requiring a connection with other individuals to multiply their species; as long as there are others which multiply in various ways without sexual intercourse, it is not justifiable to assume that those animals and plants are unmixed species and that sexual fecundity is the criterion of specific identity. Moreover, this test can hardly [if] ever have any practical value in most cases of the highest scientific interest. It is never resorted to, and, as far as I know, has never been applied with satisfactory results to settle any doubtful case. It has never assisted any anxious and conscientious naturalist in investigating the degree of relationship between closely allied animals or plants living in distant regions or in disconnected geographical areas. It will never contribute to the solution of any of those difficult cases of seeming difference or identity between extinct animals and plants found in different geological formations. In all critical cases, requiring the most minute accuracy and precision, it is discarded as unsafe and of necessity questionable. Accurate science must do without it, and the sooner it is altogether discarded, the better. But like many relics of past time, it is dragged in as a sort of theoretical bugbear and exhibited only now and then to make a false show in discussions upon the question of the unity of origin of mankind.

There is another fallacy connected with the prevailing ideas about species to which I would also allude: the fancy that species do not exist in the same way in nature as genera, families, orders, classes, and types. It is actually maintained by some that species are founded in nature in a manner different from these groups; that their existence is, as it were, more real, whilst that of the other groups is considered as ideal, even when it is admitted that these groups have themselves a natural foundation.

Let us consider this point more closely, as it involves the whole question of individuality. I wish, however, not to be understood as undervaluing the importance of sexual relations as indicative of the close ties which unite, or may unite, the individuals of the same species. I know as well as anyone to what extent they manifest themselves in nature, but I mean to insist upon the undeniable fact that these relations are not so exclusive as those naturalists would represent them who urge them as an unfailing criterion of specific identity. I would remind those who constantly forget it, that there are animals which, though specifically distinct, do unite sexually, which do produce offspring, mostly sterile, it is true, in some species, but fertile to a limited extent in others and in others even fertile to an extent which it has not yet been possible to determine. Sexual connection is the result or, rather, one of the most striking expressions of the close relationship established in the beginning between individuals of the same species, and by no means the cause of their identity in successive generations. When first created, animals of the same species paired because they were made one for the other; they did not take one another in order to build up their species, which had full existence before the first individual produced by sexual connection was born.

This view of the subject acquires greater importance in proportion as it becomes more apparent that species did not originate in single pairs, but were created in large numbers, in those numeric proportions which constitute the natural harmonies between organized beings. It alone explains the possibility of the procreation of Hybrids, as founded upon the natural relationship of individuals of closely allied species which may become fertile with one another, the more readily as they differ less, structurally.

To assume that sexual relations determine the species, it should further be shown that absolute promiscuousness of sexes among individuals of the same species is the prevailing characteristic of the animal kingdom, while the fact is that a large number even of animals, not to speak of Man, select their mate for life and rarely have any intercourse with others. It is a fact known to every farmer that different breeds of the same species are less inclined to mingle than individuals of the same breed. For my own part, I cannot conceive how moral philosophers, who urge the unity of origin of Man as one of the fundamental principles of their religion, can at the same time justify the necessity which it involves of a sexual intercourse between the nearest blood relations of that assumed first and unique human family, when such a connection is revolting even to the savage. Then again, there are innumerable species in which vast numbers of individuals are never developed sexually, others in which sexual individuals appear only now and then at remote intervals, while many intermediate generations are produced without any sexual connection, and others still which multiply more extensively by budding than by sexual generation. I need not again allude here to the phenomena of alternate generation, now so well known among Acalephs and Worms, nor to the polymorphism of many other types. Not to acknowledge the significance of such facts would amount to the absurd pretension that distinctions and definitions, introduced in our science during its infancy, are to be taken as standards for our appreciation of the phenomena in nature, instead of framing and remodelling our standards according to the laws of nature as our knowledge extends. It is, for instance, a specific character of the Horse and the Ass to be able to connect sexually with each other and thus to produce an offspring different from that which they bring forth among themselves. It is characteristic of the Mare, as the representative of its species, to bring forth a Mule with the Jackass, and of the Stallion to procreate Hinnies with the She-ass. It is equally characteristic of them to produce still other kinds of halfbreeds with the Zebra, the Dauw, etc. And yet in face of all these facts, which render sexual reproduction, or at least promiscuous intercourse among the representatives of the same species, so questionable a criterion of specific identity, there are still naturalists who would represent it as an unfailing test, only that they may sustain one single position, that all men are derived from one single pair.

These facts, with other facts which go to show more extensively every day the great probability of the independent origin of individuals of the same species in disconnected geographical areas, force us to remove from the philosophic definition of species the idea of a community of origin, and consequently also the idea of a necessary genealogical connection. The evidence that all animals have originated in large numbers is growing so strong, that the idea that every species existed in the beginning in single pairs may be said to be given up almost entirely by naturalists. Now if this is the case, sexual derivation does not constitute a necessary specific character, even though sexual connection be the natural process of their reproduction and multiplication. If we are led to admit as the beginning of each species the simultaneous origin of a large number of individuals, if the same species may originate at the same time in different localities, these first representatives of each species, at least, were not connected by sexual derivation; and as this applies equally to any first pair, this fancied test criterion of specific identity must at all events be given up, and with it goes also the pretended real existence of the species, in contradistinction from the mode of existence of genera, families, orders, classes, and types; for what really exists are individuals, not species. We may at the utmost consider individuals as representatives of species, but no one individual nor any number of individuals represent its species only without representing also at the same time, as we have seen above (Sect. I to V), its genus, its family, its order, its class, its branch.[25]

Before attempting to prove the whole of this proposition, I will first consider the characters of the individual animals. Their existence is scarcely limited as to time and space within definite and appreciable limits. No one nor all of them represent fully, at any particular time, their species; they are always only the temporary representatives of the species, inasmuch as each species exists longer in nature than any of its individuals. All the individuals of any or of all species now existing are only the successors of other individuals which have gone before, and the predecessors of the next generations; they do not constitute the species, they represent it. The species is an ideal entity, as much as the genus, the family, the order, the class, or

[25] [For a discussion of Agassiz's views on the species question, its relationship to the origin of man, and related literature on these subjects see Edward Lurie, "Louis Agassiz and the Races of Man," *Isis*, XLV (1954), 227–242.]

the type; it continues to exist while its representatives die, generation after generation. But these representatives do not simply represent what is specific in the individual, they exhibit and reproduce in the same manner, generation after generation, all that is generic in them, all that characterizes the family, the order, the class, the branch, with the same fulness, the same constancy, the same precision. Species then exist in nature in the same manner as any other groups, they are quite as ideal in their mode of existence as genera, families, etc., or quite as real. But individuals truly exist in a different way; no one of them exhibits at one time all the characteristics of the species, even though it be hermaphrodite, neither do any two represent it, even though the species be not polymorphous, for individuals have a growth, a youth, a mature age, an old age, and are bound to some limited home during their lifetime. It is true species are also limited in their existence; but for our purpose we can consider these limits as boundless, inasmuch as we have no means of fixing their duration, either for the past geological ages, or for the present period, whilst the short cycles of the life of individuals are easily measurable quantities. Now as truly as individuals, while they exist, represent their species for the time being and do not constitute them, so truly do these same individuals represent at the same time their genus, their family, their order, their class, and their type, the characters of which they bear as indelibly as those of the species.

As representatives of Species individual animals bear the closest relations to one another; they exhibit definite relations also to the surrounding elements, and their existence is limited within a definite period.

As representatives of Genera these same individuals have a definite and specific ultimate structure, identical with that of the representatives of other species.

As representatives of Families these same individuals have a definite figure exhibiting, with similar forms of other genera, or for themselves, if the family contains but one genus, a distinct specific pattern.

As representatives of Orders these same individuals stand in a definite rank when compared to the representatives of other families.

As representatives of Classes these same individuals exhibit the plan of structure of their respective type in a special manner, carried out with special means and in special ways.

As representatives of Branches these same individuals are all organized upon a distinct plan, differing from the plan of other types.

Individuals then are the bearers, for the time being, not only of specific characteristics, but of all the natural features in which animal life is displayed in all its diversity.

Viewing individuals in this light, they resume all their dignity; they are no longer absorbed in the species to be forever its representatives, without ever being anything for themselves. On the contrary, it becomes plain, from this point of view, that the individual is the worthy bearer, for the time being, of all the riches of nature's wealth of life. This view further teaches us how we may investigate, not only the species in the individual, but the genus also, the family, the order, the class, the branch, as indeed naturalists have at all times proved in practice whilst denying the possibility of it in theory.

Having thus cleared the field of what does not belong therein, it now remains for me to show what in reality constitutes species, and how they may be distinguished with precision within their natural limits.

If we would not exclude from the characteristics of species any feature which is essential to it, nor force into it any one which is not so, we must first acknowledge that it is one of the characters of species to belong to a given period in the history of our globe, and to hold definite relations to the physical conditions then prevailing and to animals and plants then existing. These relations are manifold and are exhibited: 1st, in the geographical range natural to any species, as well as in its capability of being acclimated in countries where it is not primitively found; 2d, in the connection in which they stand to the elements around them, when they inhabit either the water, or the land, deep seas, brooks, rivers and lakes, shoals, flat, sandy, muddy, or rocky coasts, limestone banks, coral reefs, swamps, meadows, fields, dry lands, salt deserts, sandy deserts, moist land, forests, shady groves, sunny hills, low regions, plains, prairies, high table-lands, mountain peaks, or the frozen barrens of the Arctics, etc.; 3d, in their dependence upon this or that kind of food for their sustenance; 4th, in the duration of their life; 5th, in the mode of their association with one another, whether living in flocks, small companies, or isolated; 6th, in the period of their reproduction; 7th, in the changes they undergo during their growth and the periodicity of these changes in their metamorphosis; 8th, in their association with other beings, which is more or less close, as it may only lead to a constant association in some, whilst in others it amounts to parasitism; 9th, specific characteristics are further exhibited in the size animals attain, in the proportions of their parts to one another, in their ornamentation, etc., and all the variations to which they are liable.

As soon as all the facts bearing upon these different points have been fully ascertained, there can remain no doubt respecting the natural limitation of species; and it is only the insatiable desire of describing new species from insufficient data which has led to the introduction in our systems of so many doubtful species, which add nothing to our real knowledge, and only go to swell the nomenclature of animals and plants already so intricate.

Assuming then that species cannot always be identified at first sight, that it may require a long time and patient investigations to ascertain their natural limits; assuming further that the features alluded to above are among the most prominent characteristics of species, we may say that species are based upon well determined relations of individuals to the world around them, to their kindred, and upon the proportions and relations of their parts to one another, as well as upon their ornamentation. Well digested descriptions of species ought, therefore, to be comparative; they ought to assume the character of biographies and attempt to trace the origin and follow the development of a species during its whole existence. Moreover, all the changes which species may undergo in course of time, especially under the fostering care of man, in the state of domesticity and cultivation, belong to the history of the species; even the anomalies and diseases to which they are subject belong to their cycle, as well as their natural variations. Among some species variation of color is frequent, others never change, some change periodically, others accidentally; some throw off certain ornamental appendages at regular times, the Deers their horns, some Birds the ornamental plumage they wear in the breeding season, etc. All this should be ascertained for each, and no species can be considered as well defined and satisfactorily characterized the whole history of which is not completed to the extent alluded to above. The practice prevailing since Linnæus, of limiting the characteristics of species to mere diagnoses, has led to the present confusion of our nomenclature, and made it often impossible to ascertain what were the species the authors of such condensed descriptions had before them. But for the tradition which has transmitted, generation after generation, the knowledge of these species among the cultivators of science in Europe, this confusion would be still greater; but for the preservation of most original collections it would be inextricable. In countries which, like America, do not enjoy these advantages, it is often hopeless to attempt critical investigations upon doubtful cases of this kind. One of our ablest and most critical investigators, the lamented Dr. Harris, has very forcibly set forth the difficulties under which American naturalists labor in this respect, in the Preface to his *Report upon the Insects Injurious to Vegetation.*[26]

SECTION VII

OTHER NATURAL DIVISIONS AMONG ANIMALS

Thus far I have considered only those kinds of divisions which are introduced in almost all our modern classifications, and attempted to show that these groups are founded in nature and ought not to be considered as artificial devices invented by man to facilitate his studies. Upon the closest scrutiny of the subject I find that these divisions cover all the categories of relationship which exist among animals, as far as their structure is concerned.

Branches or *types* are characterized by the plan of their structure;

Classes, by the manner in which that plan is executed, as far as ways and means are concerned;

Orders, by the degrees of complication of that structure;

Families, by their form, as far as determined by structure;

Genera, by the details of the execution in special parts; and

Species, by the relations of individuals to one another and to the world in which they live, as well as by the proportions of their parts, their ornamentation, etc.

[26] [Pp. 1–8.]

And yet there are other natural divisions which must be acknowledged in a natural zoological system; but these are not to be traced so uniformly in all classes as the former — they are in reality only limitations of the other kinds of divisions.

A class in which one system of organs may present a peculiar development while all the other systems coincide may be subdivided into sub-classes; for instance, the Marsupialia when contrasted with the Placental Mammalia. The characters upon which such a subdivision is founded are of the kind upon which the class itself is based, but do not extend to the whole class. An order may embrace natural groups, of a higher value than families, founded upon ordinal characters, which may yet not determine absolute superiority or inferiority, and therefore not constitute for themselves distinct orders; as the characters upon which they are founded, though of the kind which determines orders, may be so blended as to determine superiority in one respect, while with reference to some other features they may indicate inferiority. Such groups are called sub-orders. The order of Testudinata, may best illustrate this point, as it contains two natural sub-orders.[27] A natural family may exhibit such modifications of its characteristic form that upon these modifications subdivisions may be distinguished which have been called sub-families by some authors, tribes or legions by others. In a natural genus a number of species may agree more closely than others in the particulars which constitute the genus and lead to the distinction of sub-genera. The individuals of a species, occupying distinct fields of its natural geographical area, may differ somewhat from one another, and constitute varieties, etc.

These distinctions have long ago been introduced into our systems, and every practical naturalist who has made a special study of any class of the animal kingdom must have been impressed with the propriety of acknowledging a large number of subdivisions to express all the various degrees of affinity of the different members of any higher natural group. Now while I maintain that the branches, the classes, the orders, the families, the genera, and the species are groups established in nature respectively upon different categories, and while I feel prepared to trace the natural limits of these groups by the characteristic features upon which they are founded, I must con-

[27] See my *Contributions,* I, 308.

fess at the same time that I have not yet been able to discover the principle which obtains in the limitation of their respective subdivisions.[28] All I can say is that all the different categories considered above, upon which branches, classes, orders, families, genera, and species are founded, have their degrees, and upon these degrees sub-classes, sub-orders, sub-families, and sub-genera have been established. For the present these subdivisions must be left to arbitrary estimations, and we shall have to deal with them as well as we can, as long as the principles which regulate these degrees in the different kinds of groups are not ascertained. I hope nevertheless that such arbitrary estimations are forever removed from our science, as far as the categories themselves are concerned.

Thus far, inequality of weight seems to be the standard of the internal valuation of each kind of group; and this inequality extends to all groups, for even within the branches there are classes more closely related among themselves than others: Polypi and Acalephs, for instance, stand nearer to one another than to Echinoderms; Crustacea and Insects are more closely allied to one another than to Worms, etc. Upon such degrees of relationship between the classes, within their respective branches, the so-called sub-types have been founded, and these differences have occasionally been exaggerated so far as to give rise to the establishment of distinct branches. Upon similar relations between the branches, sub-kingdoms have also been distinguished, but I hardly think that such far-fetched combinations can be considered as natural groups; they seem to me rather the expression of a relation arising from the weight of their whole organization, as compared with that of other groups, than the expression of a definite relationship.

[28] Professor James D. Dana has thrown out some valuable suggestions upon this point in his review of my *Contributions*. [In this note to the 1859 edition Agassiz refers to Dana's review of the *Essay* as it appeared in the first volume of the *Contributions*. See *American Journal of Science*, XXV (2d ser., 1858), 126–128, 202–216, 321–341, especially p. 333.]

SECTION VIII

SUCCESSIVE DEVELOPMENT OF CHARACTERS

It has been repeated, again and again, that the characters distinguishing the different types of the animal kingdom were developed in the embryo in the successive order of their importance: first the structural features of their respective branches, next the characters of the class, next those of the order, next those of the family, next those of the genus, and finally those of the species. This assertion has met with no direct opposition; on the contrary, it seems to have been approved almost without discussion and to be generally taken for granted now. The importance of the subject requires, however, a closer scrutiny; for if Embryology is to lead to great improvements in Zoology, it is necessary at the outset to determine well what kind of information we may expect it to furnish to its sister science. Now I would ask if, at this day, zoologists know with sufficient precision what are typical, class, ordinal, family, generic, and specific characters, to be justified in maintaining that, in the progress of embryonic growth, the features which become successively prominent correspond to these characters and in the order of their subordination? I doubt it. I will say more: I am sure there is no such understanding about it among them, for if there was, they would already have perceived that this assumed coincidence between the subordination of natural groups among full-grown animals and the successive stages of growth during their embryonic period of life does not exist in nature. It is true there are certain features in the embryonic development which may suggest the idea of a progress from a more general typical organization to its ultimate specialization, but it nowhere proceeds in that stereotyped order of succession, nor indeed even in a general way in the manner thus assumed.

Let us see whether it is not possible to introduce more precision in this matter. Taking for granted that what I have said about the characteristics of the natural groups in the animal kingdom is correct, that we have, 1st, four great typical branches of the animal kingdom, characterized by different plans of structure; 2d, classes characterized by the ways in which and the means with which these plans of structure are executed; 3d, orders, characterized by the degrees of simplicity or complication of that structure; 4th, families, characterized by differences of form, or by the structural peculiarities determining form; 5th, genera, characterized by ultimate peculiarities of structure in the parts of the body; 6th, species, characterized by relations and proportions of parts among themselves, and of the individuals to one another and to the surrounding mediums; we reach, finally, the individuals, which for the time being represent not only the species with all their varieties, and variations of age, sex, size, etc., but also the characteristic features of all the higher groups. We have thus at one end of the series the most comprehensive categories of the structure of animals, while at the other end we meet individual beings. Individuality on one side, the most extensive divisions of the animal kingdom on the other. Now, to begin our critical examination of the progress of life in its successive manifestations with the extremes, is it not plain, from all we know of Embryology, that individualization is the first requirement of all reproduction and multiplication, and that an individual germ (or a number of them), an ovarian egg, or a bud, is first formed and becomes distinct as an individual from the body of the parent, before it assumes either the characters of its great type or those of its class, order, etc.? This fact is of great significance as showing the importance of individuality in nature. Next, it is true, we perceive generally the outlines of the plan of structure, before it becomes apparent in what manner that plan is to be carried out; the character of the type is marked out, in its most general features, before that of the class can be recognized with any degree of precision. Upon this fact we may base one of the most important generalizations in Embryology.

It has been maintained in the most general terms that the higher animals pass during their development through all the phases characteristic of the inferior classes. Put in this form, no statement can be further from the truth, and yet there are decided relations within certain limits between the embryonic stages of growth of higher animals and the permanent characters of others of an inferior grade. Now the fact mentioned above enables us to mark with precision the limits within which these relations may be traced. As eggs, in their primitive condition, animals do not differ one from the other; but as soon as the embryo has begun to show any characteristic features it presents such peculiarities as distinguish its type. It cannot therefore be said that any animal passes through phases of development, which are not included within the limits of its own type; no Vertebrate is or resembles, at any time, an Articulate, no Articulate a Mollusk, no Mollusk a Radiate, and *vice versa*. Whatever correlations between the young of higher animals and the perfect condition of inferior ones may be traced, they are always limited to representatives of the same great types; for instance, Mammalia and Birds in their earlier development exhibit certain features of the lower classes of Vertebrates, such as the Reptiles or Fishes; Insects recall the Worms in some of their earlier stages of growth, etc., but even this requires qualifications to which we shall have to refer hereafter. However, this much is already evident, that no higher animal passes through phases of development recalling all the lower types of the animal kingdom, but only such as belong to its own branch.[29] What has been said of the infusorial character of young embryos of Worms, Mollusks, and Radiates can no longer stand before a serious criticism, because, in the first place, the animals generally called Infusoria cannot themselves be considered as a natural class; and, in the second place, those to which a reference is made in this connection are themselves free-moving embryos.[30]

With the progress of growth and in proportion as the type of an animal becomes more distinctly marked in its embryonic state, the plan of structure appears also more distinctly in the peculiarities of that structure, that is to say, in the ways in which and the means by which the plan, only faintly indicated at first, is to be carried out and become prominent, and by this the class character is pointed out. For instance, a wormlike insect larva will already show by its tracheæ that it is to be an Insect and not to remain a Worm, as it at first appears to be; but the complications of that special structure, upon which the orders of the class of Insects are based, do not yet appear; this is perfected only at a late period in the embryonic life. At this stage we frequently notice already a remarkable advance of the features characteristic of the families over those characteristic of the

[29] [The preceding statements represent Agassiz's unwillingness to accept the recapitulation concept or biogenetic law in its most radical form. His criticism stemmed from the conviction that the notion of a complete recapitulation of all racial history from lower to higher forms came very close to a theory of organic evolution.]

[30] See above, Chap. I, Sect. XVIII.

order; for instance, young Hemiptera, young Orthoptera may safely be referred to their respective families from the characteristics they exhibit before they show those peculiarities which characterize them as Hemiptera or as Orthoptera; young Fishes may be known as members of their respective families before the characters of their orders are apparent, etc.

It is very obvious why this should be so. With the progress of the development of the structure the general form is gradually sketched out, and it has already reached many of its most distinctive features before all the complications of the structure which characterize the orders have become apparent; and as form characterizes essentially the families, we see here the reason why the family type may be fully stamped upon an animal before its ordinal characters are developed. Even specific characters, as far as they depend upon the proportions of parts and have on that ground an influence in modifying the form, may be recognized long before the ordinal characters are fully developed. The Snapping-Turtle, for instance, exhibits its small crosslike sternum, its long tail, its ferocious habits even before it leaves the egg, before it breathes through lungs, before its derm is ossified to form a bony shield, etc.; nay, it snaps with its gaping jaws at anything brought near, though it be still surrounded by its amnios and allantois, and its yolk still exceeds in bulk its whole body.[31] The calf assumes the form of the bull before it bears the characteristics of the hollow-horned Ruminants; the fawn exhibits all the peculiarities of its species before those of its family are unfolded.

With reference to generic characters, it may be said that they are scarcely [if] ever developed in any type of the animal kingdom before the specific features are for the most part fully sketched out, if not completely developed. Can there be any doubt that the human embryo belongs to the genus Homo, even before it has cut a tooth? Is not a kitten or a puppy distinguishable as a cat or a dog before the claws and teeth tell their genus? Is this not true also of the Lamb, the Kid, the Colt, the Rabbits, and the Mice, of most Birds, most Reptiles, most Fishes, most Insects, Mollusks and Radiates? And why

[31] Prince Maximilian von Wied-Neuwied quotes as a remarkable fact that the *Chelonara sperpentina* bites as soon as it is hatched. I have seen it snapping in the same fierce manner as it does when full grown, at a time when it is still a pale, almost colorless embryo, wrapped up in fetal envelopes, with a yolk larger than itself hanging from its sternum, three months before it is hatched.

should this be? Simply because the proportions of parts, which constitute specific characters, are recognizable before their ultimate structural development, which characterizes genera, is completed.

It seems to me that these facts are likely to influence the future progress of Zoology in enabling us gradually to unravel more and more distinctly the features which characterize the different subordinate groups of the animal kingdom. The views I have expressed above of the respective value and the prominent characteristics of these different groups have stood so completely the test in this analysis of their successive appearance, that I consider this circumstance as adding to the probability of their correctness.

But this has another very important bearing, to which I have already alluded in the beginning of these remarks. Before Embryology can furnish the means of settling some of the most perplexing problems in Zoology, it is indispensable to ascertain first what are typical, classic, ordinal, family, generic, and specific characters; and as long as it could be supposed that these characters appear necessarily during the embryonic growth in the order of their subordination, there was no possibility of deriving from embryological monographs that information upon this point, so much needed in Zoology and so seldom alluded to by embryologists. Again, without knowing what constitutes truly the characters of the groups named above, there is no possibility of finding out the true characters of a genus of which only one species is known, of a family which contains only one genus, etc., and for the same reason no possibility of arriving at congruent results with reference to the natural limitations of genera, families, orders, etc., without which we cannot even begin to build up a permanent classification of the animal kingdom; and, still less, hope to establish a solid basis for a general comparison between the animals now living and those which have peopled the surface of our globe in past geological ages.

It is not accidentally I have been led to these investigations, but by necessity. As often as I tried to compare higher or more limited groups of animals of the present period with those of former ages, or early stages of growth of higher living animals with full-grown ones of lower types, I was constantly stopped in my progress by doubts as to the equality of the standards I was applying, until I made the standards themselves the object of direct and very extensive investigations, covering indeed a much wider ground than would appear from these remarks; for, upon these principles, I have already remodelled, for my own convenience, nearly the whole animal kingdom and introduced in almost every class very unexpected changes in the classification.

I have already expressed above my conviction that the only true system is that which exists in nature, and as therefore no one should have the ambition of erecting a system of his own, I will not even attempt now to present these results in the shape of a diagram, but remain satisfied to express my belief that all we can really do is, at best, to offer imperfect translations in human language of the profound thoughts, the innumerable relations, the unfathomable meaning of the plan actually manifested in the natural objects themselves; and I should consider it as my highest reward should I find, after a number of years, that I had helped others on in the right path.

SECTION IX [32]

THE CATEGORIES OF ANALOGY

Thus far we have considered those relations only among animals which are founded upon strictly homological features of their structure. We now proceed to examine the more remote and less definite relations, which are called analogies.

It has already been stated in what way homologies differ from analogies. Homology is that kind of relationship which is founded upon identity of structure in different animals belonging to natural divisions of the same kind; while analogy is a resemblance arising from the combination of features characteristic of one natural group with those of another group.[33] We have indeed seen that all the animals belonging to the same branch are homologous, as far as the plan of their structure is concerned; that all the members of the same class are homologous, as far as the mode of execution of that structure

[32] [This entire section was added in the 1859 edition.]

[33] Homology has also been defined as the relationship arising from identity of structure without reference to function, while analogy is based upon similarity of function, without reference to structure. The definition given above is more precise, as it embraces all the different categories of analogy and homology

is concerned; that all the members of the same order are homologous in the complication of their structure; that all the representatives of the same family are homologous in form; that the different genera of one and the same family exhibit homologous peculiarities in the details of their structure; and that even within the narrow limits of species we may still trace homologous features, among the genera which have numerous representatives, even when such resemblances do not extend to the species of closely allied genera. It is plain from this that the categories of homology are as numerous and diversified as the essential kinds of differences which we may trace in the structure of animals; or, in other words, we have branch homologies, class homologies, ordinal homologies, family homologies, and specific homologies. Examples of the more comprehensive kinds of these homologies will occur to every practical geologist. As to specific homologies, they are particularly traceable in those structural features which determine the proportions among the parts; as, for instance, when all the species of one genus are either long-necked, short-tailed, long-legged, etc., while those of closely allied genera may present reverse proportions, etc.

Let us now see what are the categories of analogy and how far it is possible, under all circumstances, to distinguish homological and analogical features. If analogy is a resemblance arising from a combination of features characteristic of one group with those characteristic of another group (such as class characters of one class with those of another class, or those of families of another class), then the investigation will only require the recognition of the different categories of structure already considered (such as branches, classes, orders, etc.), and a correct appreciation of the mode of their combination with those of another group. It will, for instance, be sufficient to ascertain in what manner the features resulting from a certain mode of execution of the homologies of one type are combined with structures of another type; or, in other words, to recognize any feature wherever it appears, and not merely within the limits within which structures are strictly homologous. The study of analogies is therefore limited to the investigation of more or less distinct features that are naturally characteristic of one kind of group, in their combination with features of groups of another kind. For instance, the similarity between an insect wing and the wing of a bird is based upon analogy. The entire difference of structure between the organs of flight in these two classes of animals forbids our considering the resemblance which exists between them as homological, for they are not built upon homologous structures. But there is analogy between them, inasmuch as the peculiar structure characteristic of two different types is worked up into organs that appear the same because they perform similar functions.

Admitting these distinctions to be correct, the categories of analogy must be like those of homology; either analogies of branch, or of class, or of order, or of family, or of genus, or of species; and these analogies may either be observed between different branches, classes, orders, families, genera, and species; or features characteristic of branch or of class may be limited to certain families, or even to genera of other branches and other classes; so that the study of analogies becomes very difficult and highly complicated; and these complications have, no doubt, been the source of most errors and inaccuracies in the attempts that have been made to classify the animal kingdom.

Branch analogies. The plan[s] of structure characteristic of the four branches of the animal kingdom are so peculiar that we nowhere find analogies of this kind extending from one branch to all the representatives of another branch. On the contrary, they extend generally to minor divisions of some classes, and rarely to entire classes. Yet among Mollusks all the Cephalopods have some analogy with the Radiates in the arrangement of their arms around the mouth. All the Bryozoa have a striking analogy with the Polyps in the crown which spreads around their upper part; and so it is with the tentacles of a large number of the Dorsibranchiate Annelids. There is an unmistakable analogy between the structure of the solid frame of Echinoderms (especially in the star-fishes) and the plan of structure of the Articulates; so much so that Oken does not hesitate to refer the Echinoderms to the type of Articulates, mistaking their analogy for true homology.

Class analogies. The ways in which and the means by which the plan of structure of one class is carried out, as compared with another class, frequently produce striking analogies. For instance, among Vertebrates the whole class of Birds is winged; and wings constructed like the wing of Birds exist in no other class. Yet the bats are also winged; and many Fishes which are capable of rising above the water are also described as winged. But the wing of a bat is homologous to the foreleg of the other Mammalia; and only analogous to that of Birds; for it exhibits the special homologies of the class of Mammalia, and not those of the class of Birds.[34] The same is true of the so-called wings of the flying fishes, in which the wing is a fin, homologous to the pectoral fin of other bony fishes, and not constructed in the same way as the wing of the bat or that of the bird. The wing of Insects is entirely different, and its analogy with the wing of birds more remote than that of the bat and of the flying fish, inasmuch as it is not an analogy between members of different classes of the same branch, but between two classes of different branches, differing therefrom in the plan of structure, and not only in the mode of execution of one and the same plan.

Ordinal analogies. As orders are founded upon the complications of the structure which characterizes the different classes, it is not likely that ordinal analogies will occur between the different orders of one and the same class; we may rather expect them to be prominent between the orders of closely allied classes or between the orders of a higher class and the lower classes of the same branch. We find, for instance, a remarkable correspondence between the orders of the class of Batrachians, and those of the class of true Reptiles.[35] The same may be said of the order of Cetacea in the class of Mammalia, as compared to the whole class of Fishes, or of the lower order of the Insects (the Myriapods) as compared to the class of Worms, or of the lower order of Acalephs (the Hydroids) as compared to the class of Polyps.[36] An accurate knowledge of these kinds of analogies is of the utmost importance for the study of the true affinities of animals, since a misapprehension of the real value of their structural features has again and again misled zoologists into combining such groups as if they were truly related. In the beginning of the last century, for instance, the Cetacea were generally united with the Fishes, to which they are only analogous; and even to this day we see the Hydroids,

[34] As limbs of Vertebrates these two kinds of wings are, nevertheless, homologous; but as wings they are only analogous.

[35] For further details see the second Part of the first volume of my *Contributions* (1857), Sect. III, pp. 252–255.

[36] For further details I must refer to the third volume of my *Contributions* now in the press [Boston, 1860.]

which are true Acalephs of a lower order, united into one class with the Polypi.

Family analogies. It requires little familiarity with the animal kingdom to know how strong may be the resemblance between the forms of animals, even when they belong to entirely different types; but unless their pattern be determined by identical structural features, their form certainly cannot be considered as homologous; and however close the resemblance may be externally, an attempt to distinguish between analogical and homological forms cannot fail to add precision to our zoological investigations. When, for instance, the form of the Worms is compared with that of the Holothurians, it should be borne in mind that in the Worms, according to the plan of their structure and their homology to the other Articulates, their longer diameter is the longitudinal diameter; while the longer diameter of the Holothurians, when identified by their homologies with the other Radiates, is their vertical diameter. This shows at once, that however similar to one another, the form of the Holothurians is only analogous to that of the Worms.

The limits within which similar forms may be homologous appear to be very wide and to extend beyond the limits of their respective classes. The form of the Salamanders and the Lizards, for instance, is certainly homological, though they are members of different classes; yet similar forms within the same class are not necessarily homologous — for instance, the long snout of Syngnathus and that of Fistularia, or the flat heads of Lophius and of Scaphirhynchus are only remotely analogous, their structure being entirely different. The forms of animals have been so imperfectly studied and the structural elements which determine them so little considered, that the time has hardly come yet to determine with any degree of accuracy the analogies and homologies of the form of animals. Considered with reference to their position, the six pairs of articulated appendages which are placed upon the sides of the mouth of the horse-shoe crab (Limulus) are truly homologous to the jaws of the higher Crustacea; but by their form they resemble the thoracic legs of the latter; and yet, as appendages to the normal rings of an Articulate, all these parts are homologous. Here therefore it becomes necessary to remember that while the appendages of the mouth of Limulus are only analogous to the legs of the Decapods, as far as their form is concerned, these organs are yet homologous as parts of the body of an Articulate. This and similar cases may show how wide a field of investigation lies before us in the study and discrimination of homological and analogical forms.

Generic analogies. As the generic characters are based upon peculiarities of structure limited to some part or other of an animal, we may expect to find the generic analogies reduced to a resemblance of certain parts of the body and not extending to its general appearance. For while genera, as members of a family, must exhibit the same form, combined with the structural complication of their order, it is obvious that if there is any generic analogy between animals of different families their whole form may be widely different and the complication of their structure exhibit entirely different combinations, or be based upon different modes of execution, if they belong to different classes, and even be constructed upon different plans of structure if they belong to different branches; and yet some of their parts should be similar in some way or other, in order to present a generic analogy.

Now such generic analogies are rather frequent and may be traced between animals of widely different families belonging to different orders, nay, even to different classes and to different branches; for instance, there is a marked generic analogy between the dentition of the Insectivora, of the class of Mammalia, and that of the Characini of the class of Fishes, so also between some genera of the family of Sparoids and those of the Chromids, between some genera of the family of Insectivora and the Rodentia, and between some of the family of Bombryces and of the Papiliones, etc.

Specific analogies. If the characteristic features of species be truly found in the relations which animals bear to the surrounding world or to one another, and in the relative proportions of their parts and their ornamentation, we cannot fail to find specific analogies resulting from these different aspects in animals belonging to different genera, to different families, to different orders, and even to different classes and branches. As far as they are aquatic, animals belonging to different genera which number terrestrial species also have a certain analogy with one another. All animals living in pairs or in flocks, or isolated, may in this respect be considered as having an analogy to one another, especially if they belong to genera in which different species bear these different relations to one another. But it is in the proportions of the parts to one another in the species of different genera belonging to the same family or even to different families of the same class, and in the ornamentation of their surface, that we observe the most numerous specific analogies. Reference has already been made to the specific homologies resulting from the relative length of the head, the neck, the tail, etc. But there is a specific analogy only between the Zerda, a species of dog found in the interior of Africa, which is characterized by the extraordinary length of its ears, and those species of hare which live also in the desert and have much larger and longer ears than those inhabiting the woods and marshes. This analogy is no doubt owing to the fact that under the conditions in which these animals are placed they require a keener perception of sound, and yet they belong to different orders, though of the same class. This is therefore a specific analogy. The pattern of coloration may also exhibit specific analogy, as, for instance, in the transverse bands of the tiger when compared to the Quagga, in the spots of the Leopard and the Giraffe, which is so striking as to have suggested the name of the latter, Camelo-pardalis.

As it is not my intention here to trace all these analogies throughout the Animal Kingdom, these few examples may suffice to call attention to the subject and to lead hereafter to a more careful investigation of the different categories of analogy. A few more remarks may, however, find a place here to show how to distinguish analogical from homological features. As homologies, whether extensive or limited, are strictly confined within groups of the same kind, it is evident that unless any feature observed in any animal be common to all the representatives of the group in which it occurs, we shall have good reason to suspect that it is not based upon strict homology, but rather belongs to some category of analogy. If, for instance, the dorsal cord is a fundamental feature of Vertebrates, any structure in the longitudinal axis of an animal which is not structurally identical with the dorsal cord cannot be homologous with it but must be something only analogous to it; for instance, the medial stripe which appears during the early development of the embryo of the earlier Crustacea. For the farther progress of the formation of the backbone we trace the formation of arches below as well as above the dorsal cord, while in Crustacea there is a similar development only on one side. We are therefore compelled to consider the solid arches of Crustacea only as analogous structures to Vertebræ and not as homologous with them, the more so since these arches enclose not only the nervous system, as in Vertebrates, but all the other viscera besides. The system of articulation in Articulates exhibits, therefore, a branch analogy with the vertebral system of the Vertebrates, but there is no true homology between them. The class of Fishes is eminently characterized by the presence of gills, and so have Crustacea gills, and so also the Cephalopods, a large number of Gasteropods, and most Acephala. But the structure of these gills is widely different in these different classes, and their presence only constitutes class analogies and is no indication of a real affinity; while the so-called lungs of the land Gasteropods have the closest structural resemblance to the gills of the other Mollusks, thus showing a real affinity between them, while their air sacs, on account of their gill-like structure, constitute only an analogy between them and the other air-breathing animals. We may go on testing in this way the analogies and homologies in all their degrees and combinations throughout the animal kingdom and be sure to arrive at satisfactory results, provided we remember that analogies are features of one group combined with the characteristic features of another group, and not, like homologies, circumscribed within one and the same group.

SECTION X

CONCLUSIONS

The importance of such an investigation as the preceding must be obvious to every philosophical investigator. As soon as it is understood that all the different groups introduced into a natural system may have a definite meaning; as soon as it can be shown that each exhibits a definite relation among living beings, founded in nature and no more subject to arbitrary modifications than any other law expressing natural phenomena; as soon as it is made plain that the natural limits of all these groups may be ascertained by careful investigations, the interest in the study of classification or the systematic relationship existing among all organized beings, which has almost ceased to engage the attention of the more careful original investigators, will be revived, and the manifold ties which link together all animals and plants as the living expression of a gigantic conception, carried out in the course of time, like a soul-breathing epos, will be scrutinized anew, determined with greater precision, and expressed with increasing clearness and propriety. Fanciful and artificial classifications will gradually lose their hold upon a better informed community; scientific men themselves will be restrained from bringing forward immature and premature investigations; no characteristics of new species will have a claim upon the notice of the learned which has not been fully investigated and compared with those most closely allied to it; no genus will be admitted, the structural peculiarities of which are not clearly and distinctly illustrated; no family will be considered as well founded which shall not exhibit a distinct system of forms intimately combined and determined by structural relations; no order will appear admissible which shall not represent a well-marked degree of structural complication; no class will deserve that name which shall not appear as a distinct and independent expression of some general plan of structure, carried out in a peculiar way and with peculiar means; no type will be recognized as one of the fundamental groups of the animal kingdom which shall not exhibit a plan of its own, not convertible into another. No naturalist will be justified in introducing any one of these groups into our systems without showing: 1st, that it is a natural group; 2d, that it is a group of this or that kind, to avoid, henceforth, calling groups that may be genera, families; groups that may be orders, families; groups that may be orders or classes, classes or branches, respectively; 3d, that the characters by which these groups may be recognized are in fact respectively specific, generic, family, ordinal, classic, or typical characters, so that our works shall no longer exhibit the annoying confusion which is to be met almost everywhere, of generic characters in the diagnoses of species, or of family and ordinal characters in the characteristics of classes and branches.[37]

It may perhaps be said that all this will not render the study of

[37] As I do not wish to be personal, I will refrain from quoting examples to justify this assertion. I would only request those who care to be accurate to examine critically almost any description of species, any characterization of genera, of families, of orders, of classes, and of types, to satisfy themselves that characters of the same kind are introduced almost indiscriminately to distinguish all these groups.

Zoology more easy. I do not expect that it will; but if an attentive consideration of what I have stated in the preceding pages respecting classification should lead to a more accurate investigation of all the different relations existing among animals, and between them and the world in which they live, I shall consider myself as having fully succeeded in the object I have had in view from the beginning in this inquiry. Moreover, it is high time that certain zoologists who would call themselves investigators should remember that natural objects, to be fully understood, require more than a passing glance;[38] they should imitate the example of astronomers, who have not become tired of looking into the relations of the few members of our solar system to determine, with increased precision, their motions, their size, their physical constitution, and keep in mind that every organized being, however simple in its structure, presents to our appreciation far more complicated phenomena within our reach than all the celestial bodies put together; they should remember that as the great literary productions of past ages attract ever anew the attention of scholars who can never feel that they have exhausted the inquiry into their depth and beauty, so the living works of God, which it is the proper sphere of Zoology to study, would never cease to present new attractions to them, should they proceed to the investigation with the right spirit. Their studies ought indeed inspire everyone with due reverence and admiration for such wonderful productions.

The subject of classification in particular, which seems to embrace apparently so limited a field in the science of animals, cannot be rightly and fully understood without a comprehensive knowledge of all the topics alluded to in the preceding pages.

[38] The mere indication of the existence of a species is a poor addition to our knowledge, when compared to those monographs in which either the structure or the development of a single animal is fully illustrated, such as Owen, *Memoir on the Pearly Nautilus*. It may, indeed, be said that there hardly appears one such work every other year, and that thousands of years will be required, at the present rate of our progress, to investigate satisfactorily and in all their relations the hundred thousands of living and extinct animals now known to exist. It might afford some consolation to those impatient spirits who quarrel with their fellow-students about the discovery of a hair upon a stuffed skin, if they only knew what rich harvests remain to be gathered.

CHAPTER III

NOTICE OF THE PRINCIPAL SYSTEMS OF ZOOLOGY

SECTION I

GENERAL REMARKS UPON MODERN SYSTEMS

WITHOUT attempting to give an historical account of the leading features of all zoological systems, it is proper that I should here compare critically the practice of modern naturalists with the principles discussed above. With this view it would hardly be necessary to go back beyond the publication of the *Animal Kingdom,* by Cuvier, were it not that Cuvier is still represented by many naturalists, and especially by Ehrenberg,[1] and some other German zoologists, as favoring the division of the whole animal kingdom into two great groups, one containing the Vertebrates, and the other all the remaining classes, under the name of Invertebrates, while in reality it was he who first, dismissing his own earlier views, introduced into the classification of the animal kingdom that fourfold division which has been the basis of all improvements in modern Zoology. He first showed that animals differ, not only by modifications of one and the same organic structure, but are constructed upon four different plans of structure, forming natural, distinct groups, which he called Radiata, Articulata, Mollusca, and Vertebrata.

It is true that the further subdivisions of these leading groups have undergone many changes since the publication of the *Règne animal.* Many smaller groups, even entire classes, have been removed from one of his "embranchments" to another; but it is equally true that the characteristic idea which lies at the bottom of these great divisions was first recognized by him, the greatest zoologist of all time.

[1] *Die Corallenthiere des rothen Meeres* (Berlin, 1834).

The question which I would examine here in particular, is not whether the circumscription of these great groups was accurately defined by Cuvier, whether the minor groups referred to them truly belong there or elsewhere, nor how far these divisions may be improved within their respective limits, but whether there are four great fundamental groups in the animal kingdom, based upon four different plans of structure, and neither more nor less than four. This question is very seasonable, since modern zoologists, and especially Siebold, Leuckart, and Vogt have proposed combinations of the classes of the animal kingdom into higher groups, differing essentially from those of Cuvier. It is but justice to Leuckart to say that he has exhibited, in the discussion of this subject, an acquaintance with the whole range of Invertebrata,[2] which demands a careful consideration of the changes he proposes, as they are based upon a critical discrimination of differences of great value, though I think he overrates their importance. The modifications introduced by Vogt, on the contrary, appear to me to be based upon entirely unphysio-

[2] *Ueber die Morphologie . . . der wirbellosen Thiere* (1848). The readiness with which the German naturalists have acquiesced in the proposition of Leuckart to unite the Polyps and Acalephs into one class, seems to be owing to the circumstance that their opportunities for studying the Polyps have been chiefly limited to the Actiniæ. Had they been able to extend their investigations to the Astræans and Madrepores, and to the many types of Halcyonoids which characterize the Faunæ of the tropics, they could not have failed to perceive that the Polyps constitute for themselves a distinct class, founded upon a special mode of execution of the plan which distinguishes the Radiata from the other branches of the animal kingdom. Their investigations have truly shown, what several French naturalists have long maintained, that many families of Radiata, long referred to the class of Polyps, such as the Hydroids, cannot be separated from the Acalephs; but they have been misled, by the evidence thus obtained, to an exaggeration of the affinities of the Acalephs and Polyps. The Polyps, as a class, differ from the Acalephs in exhibiting radiating partitions, projecting inward from the outer wall of the body into the main cavity, and in having a digestive cavity derived from the inversion of the upper part of that wall into the upper part of the main cavity. In Acalephs there are no radiating partitions, and the digestive cavity is hollowed out of the mass of the body; the central prolongation of the body rising above the digestive cavity in the shape of oral appendages, which are never hollow as the tentacles of the Polyps are. The mouth tentacles of Cerianthus, which are hollow, are not homologous to the oral appendages of the Acalephs, but constitute only an inner row of tentacles, of the same kind as those that project around the upper margin of the main cavity. Again, the marginal tentacles of the Acalephs are homologous to those of the Polyps, while their oral appendages are characteristic of their class. I may add also that the radiating partitions of the Rugosa, which I refer to the Acalephs, as well as the Tabulata, are not homologous to the radiating partitions of the Actinoids and Halcyonoids, but correspond to the ridges of the stem of certain Halcyonoids, and are, like them, a foot secretion.

logical principles, though seemingly borrowed from that all important guide, Embryology.

The divisions adopted by Leuckart are: Protozoa (though he does not enter upon an elaborate consideration of that group), Coelenterata, Echinodermata, Vermes, Arthropoda, Mollusca, and Vertebrata. The classification adopted many years before by Siebold in his text-book of comparative anatomy is nearly the same, except that Mollusks follow the Worms, that Coelenterata and Echinoderms are united into one group, and that the Bryozoa are left among the Polyps.

Here we have a real improvement upon the classification of Cuvier, inasmuch as the Worms are removed from among the Radiates and brought nearer the Arthropods, an improvement however, which, so far as it is correct, has already been anticipated by many naturalists, since Blainville and other zoologists long ago felt the impropriety of allowing them to remain among Radiates and have been induced to associate them more or less closely with Articulates. But I believe the union of Bryozoa and Rotifera with the Worms, proposed by Leuckart, to be a great mistake; as to the separation of Coelenterata from Echinoderms, I consider it as an exaggeration of the difference which exists between Polyps and Acalephs on the one hand, and Echinoderms on the other.

The fundamental groups adopted by Vogt[3] are: Protozoa, Radiata, Vermes, Mollusca, Cephalopoda, Articulata, and Vertebrata, an arrangement which is based solely upon the relations of the embryo to the yolk, or the absence of eggs. But as I have already stated, this is an entirely unphysiological principle, inasmuch as it assumes a contrast between the yolk and the embryo, within limits which do not exist in nature. The Mammalia, for instance, which are placed, like all other Vertebrata, in the category of the animals in which there is an opposition between the embryo and the yolk, are as much formed of the whole yolk as the Echinoderms or Mollusks. The yolk undergoes a complete segmentation in Mammalia, as well as in Radiates or Worms, and most Mollusks; and the embryo when it makes its appearance no more stands out from the yolk than the little Starfish

[3] Carl Vogt, *Zoologische Briefe. Naturgeschichte der lebenden und untergegangenen Thiere* (2 vols., Frankfurt a. M., 1851), I, 70.

stands out from its yolk. These simple facts, known since Sars and Bischoff published their first observations, twenty years ago, is in itself sufficient to show that the whole principle of classification of Vogt is radically wrong.

Respecting the assertion that neither Infusoria nor Rhizopoda produce any eggs, I shall have more to say presently. As to the arrangement of the leading groups, Vertebrata, Articulata, Cephalopoda, Mollusca, Vermes, Radiata, and Protozoa in Vogt's system, it must be apparent to every zoologist conversant with the natural affinities of animals that a classification which interposes the whole series of Mollusks between the types of Articulata and Worms cannot be correct. A classification based like this, solely upon the changes which the yolk undergoes, is not likely to be the natural expression of the manifold relations existing between all animals. Indeed, no system can be true to nature which is based upon the consideration of a single part or a single organ.

After these general remarks, I have only to show more in detail why I believe that there are only four great fundamental groups in the animal kingdom, neither more nor less.

With reference to Protozoa, first, it must be acknowledged that, notwithstanding the extensive investigation of modern writers upon Infusoria and Rhizopoda, the true nature of these beings is still very little known. The Rhizopoda have been wandering from one end of the series of Invertebrata to the other, without finding a place generally acknowledged as expressing their true affinities. The attempt to separate them from all the classes with which they have been so long associated and to place them with the Infusoria in one distinct branch appears to me as mistaken as any of the former arrangements, for I do not even consider that their animal nature is yet proved beyond a doubt, though I have myself once suggested the possibility of a definite relation between them and the lowest Gasteropods.[4] Since it has been satisfactorily ascertained that the Corallines are genuine Algæ, which contain more or less lime in their structure, and since there is hardly any group among the lower animals and lower plants which does not contain simple locomotive individuals as well as compound communities, either free or adhering to the soil, I do not see that the facts known at

[4] Compare Chap. I, Sect. XVIII.

present preclude the possibility of an association of the Rhizopods with the Algæ. This would almost seem natural, when we consider that the vesicles of many Fuci contain a viscid, filamentous substance, so similar to that protruded from the body of the Rhizopods, that the most careful microscopic examination does not disclose the slightest difference in its structure from that which mainly forms the body of Rhizopods. The discovery by Schultze[5] of what he considers as the germinal granules of these beings by no means settles this question, though we have similar ovoid masses in Algæ, and though, among the latter, locomotive forms are also very numerous.

With reference to the Infusoria, I have long since expressed my conviction that they are an unnatural combination of the most heterogeneous beings. A large number of them, the Desmidieæ and Volvocinæ, are locomotive Algæ. Indeed, recent investigations seem to have established beyond all question, the fact that all the Infusoria Anentera of Ehrenberg are Algæ. The Enterodela, however, are true animals, but belong to two very distinct types, for the Vorticellidæ differ entirely from all others. Indeed, they are in my opinion the only independent animals of that group, and so far from having any natural affinity with the other Enterodela, I do not doubt that their true place is by the side of Bryozoa, among Mollusks, as I shall attempt to show presently. Isolated observations which I have been able to make upon Paramecium, Opalina, and the like, seem to me sufficient to justify the assumption that they disclose the true nature of the bulk of this group. I have seen, for instance, a Planaria lay eggs out of which Paramecium were born, which underwent all the changes these animals are known to undergo up to the time of their contraction into a chrysalis state; while the Opalina is hatched from Distoma eggs. I shall publish the details of these observations on another occasion. But if it can be shown that two such types as Paramecium and Opalina are the progeny of Worms, it seems to me to follow that all the Enterodela, with the exception of the Vorticellidæ, must be considered as the embryonic condition of that host of Worms, both parasitic and free, the metamorphosis of which is still unstudied. In this connection I might further remark that the time is not long past when Cercaria was also considered as belonging to the class of Infusoria, though at present no one doubts that

[5] Maximilian S. Schultze, *Ueber den Organismus der Polythalamien* (Leipzig, 1854).

it belongs to the cycle of Distoma; and the only link in the metamorphosis of that genus which was not known is now supplied, since, as I have stated above, the embryo which is hatched from the egg laid by the perfect Distoma is found to be Opalina.

All this leads to the conclusion, that a division of the animal kingdom to be called Protozoa, differing from all other animals in producing no eggs, does not exist in nature, and that the beings which have been referred to it have now to be divided and scattered, partly among plants, in the class of Algæ, and partly among animals, in the classes of Acephala (Vorticellæ) of Worms (Paramecium and Opalina), and of Crustacea (Rotifera); Vorticellæ being genuine Bryozoa and therefore Acephalous Mollusks, while the beautiful investigations of Dana and Leydig have proved the Rotifera to be genuine Crustacea, and not Worms.

The great type of Radiata, taking its leading features only, was first recognized by Cuvier, though he associated with it many animals which do not properly belong to it. This arose partly from the imperfect knowledge of those animals at the time, but partly also from the fact that he allowed himself in this instance to deviate from his own principle of classification, according to which types are founded upon special plans of structure. With reference to Radiata, he departed indeed from this view so far as to admit, besides the consideration of their peculiar plan, the element of simplicity of their structure as an essential feature in the typical character of these animals, in consequence of which he introduced five classes among Radiata: the Echinoderms, Intestinal Worms, Acalephs, Polypi, and Infusoria. In opposition to this unnatural association I need not repeat here what I have already stated of the Infusoria when considering the case of Protozoa; neither is it necessary to urge again the propriety of removing the Worms from among Radiata and connecting them with Articulata. There would thus remain only three classes among Radiates, — Polypi, Acalephs, and Echinoderms, — which, in my opinion, constitute really three natural classes in this great division, inasmuch as they exhibit the three different ways in which the characteristic plan of the type, radiation, is carried out in distinct structures.

Since it can be shown that Echinoderms are, in a general way, homologous in their structure with Acalephs and Polypi, it must be admitted that these classes belong to one and the same great type and that they are the only representatives of the branch of Radiata, assuming of course that Bryozoa, Corallinæ, Sponges, and all other foreign admixtures have been removed from among Polyps. Now it is this Cuvierian type of Radiata, thus freed of all its heterogeneous elements, which Leuckart undertakes to divide into two branches, each of which he considers coequal with Worms, Articulates, Mollusks, and Vertebrates. He was undoubtedly led to this exaggeration of the difference existing between Echinoderms on one side and Acalephs and Polypi on the other, by the apparently greater resemblance of Medusæ and Polypi,[6] and perhaps still more by the fact that so many genuine Acalephs, such as the Hydroids, including Tubularia, Sertularia, Campanularia, etc., are still comprised by most zoologists in the class of Polypi.

But since the admirable investigations of J. Müller have made us familiar with the extraordinary metamorphosis of Echinoderms, and since the Ctenophoræ and the Siphonophoræ have also been more carefully studied by Grube,[7] Leuckart, Kölliker, Vogt, Gegenbaur,[8] and myself, the distance which seemed to separate Echinoderms from Acalephs disappears entirely, for it is no exaggeration to say that were the Pluteus-like forms of Echinoderms not known to be an early stage in the transformation of Echinoderms, they would find as natural a place among Ctenophoræ, as the larvæ of Insects among Worms. I therefore maintain that Polypi, Acalephs, and Echinoderms constitute one indivisible primary group of the animal kingdom. The Polypoid character of young Medusæ proves this as plainly as the Medusoid character of young Echinoderms.

Further, nothing can be more unnatural than the transfer of Ctenophoræ to the type of Mollusks which Vogt has proposed, for Ctenophoræ exhibit the closest homology with the other Medusæ, as I have shown in my paper on the Beroid Medusæ of Massachusetts. The Ctenophoroid character of young Echinoderms establishes a second connection between Ctenophoræ and the other Radiata, of as great importance as the first. We have thus an anatomical link to

[6] We see here clearly how the consideration of anatomical differences which characterize classes has overriden the primary feature of branches, their plan, to exalt a class to the rank of a branch.

[7] [Adolf E. Grube, 1812–1880.]

[8] [Carl Gegenbaur, 1826–1903.]

connect the Ctenophoræ with the genuine Medusæ, and an embryological link to connect them with the Echinoderms.

The classification of Radiata may therefore stand thus: —

1st Class: *Polypi;* including two orders, the Actinoids and the Halcyonoids, as limited by Dana.

2d Class: *Acalephæ;* with the following orders: Hydroids including Siphonophoræ) Discophoræ, and Ctenophoræ.

3d Class: *Echinoderms;* with Crinoids, Asteroids, Echinoids, and Holothurioids, as orders.

The natural limits of the branch of Mollusks are easily determined. Since the Cirripeds have been removed to the branch of Articulata, naturalists have generally agreed to consider with Cuvier the Cephalopods, Pteropods, Gasteropods, and Acephala as forming the bulk of this type, and the discrepancies between modern investigators have mainly resulted from the views they have taken respecting the Bryozoa, which some consider still as Polyps, while others would unite them with the Worms, though their affinity with the Mollusks seems to me to have been clearly demonstrated by the investigations of Milne-Edwards. Vogt is the only naturalist who considers the Cephalopoda "as built upon a plan entirely peculiar,[9] though he does not show in what this peculiarity of plan consists, but only mentions the well-known anatomical differences which distinguish them from the other classes of the branch of Mollusks. These differences, however, constitute only class characters and exhibit in no way a different plan. It is indeed by no means difficult to homologize all the systems of organs of the Cephalopods with those of the other Mollusks, and with this evidence the proof is also furnished that the Cephalopods constitute only a class among the Mollusks.

As to the differences in the development of the Cephalopods and the other Mollusks, the type of Vertebrata teaches us that partial and total segmentation of the yolk are not inconsistent with unity of type, as the eggs of Mammalia and Cyclostomata undergo a total segmentation, while the process of segmentation is more or less limited in the other classes. In Birds, Reptiles, and Selachians the segmentation is only superficial; in Batrachians and most Fishes it is much deeper; and yet no one would venture to separate the Vertebrata into several distinct branches on that account. With reference

[9] *Zoologische Briefe,* I, 361.

to Bryozoa there can be no doubt that their association with Polypi or with Worms is contrary to their natural affinities. The plan of their structure is in no way radiate; it is, on the contrary, distinctly and essentially bilateral; and as soon as their close affinities with the Brachiopods, alluded to above, are fully understood, no doubt will remain of their true relation to Mollusks. As it is not within the limits of my plan to illustrate here the characters of all the classes of the animal kingdom, I will only state further, that the branch of Mollusks appears to me to contain only three classes, as follows: —

1st Class: *Acephala;* with four orders, Bryozoa, including the Vorticellæ, Brachiopods, Tunicata, and Lamellibranchiata.

2d Class: *Gasteropoda;* with three orders, Pteropoda, Heteropoda, and Gasteropoda proper.

3d Class: *Cephalopoda;* with two orders, Tetrabranchiata and Dibranchiata.

The most objectionable modification introduced in the general classification of the animal kingdom since the appearance of Cuvier's *Règne animal* seems to me to be the establishment of a distinct branch, now very generally admitted under the name of VERMES, including the Annulata, the Helminths, the Rotifera, and, as Leuckart would have it, the Bryozoa also. It was certainly an improvement upon Cuvier's system to remove the Helminths from the type of Radiates, but it was at the same time as truly a retrograde step to separate the Annelides from the branch of Articulata. The most minute comparison does not lead to the discovery of a distinct plan of structure uniting all these animals into one natural primary group. What holds them together and keeps them at a distance from other groups is not a common plan of structure, but a greater simplicity in their organization. In bringing these animals together naturalists make again the same mistake which Cuvier committed when he associated the Helminths with the Radiates, only in another way and upon a greater scale. The Bryozoa are, as it were, depauperated Mollusks, as Aphanes and Alchemilla are depauperated Rosaceæ. Rotifera are in the same sense the lowest Crustacea; while Helminths and Annelides constitute together the lowest class of Articulata. This class is connected by the closest homology with the larval states of Insects; the plan of their structure is identical, and there exists between them only such structural differences as constitute classes. Moreover, the Helminths are linked to the Annelides in the same manner as the apodal larvæ of Insects are to the most highly organized caterpillars. It may truly be said that the class of Worms represents, in perfect animals, the embryonic states of the higher Articulata. The two other classes of this branch are the Crustacea and the Insects, respecting the limits of which as much has already been said above as is necessary to state here.

The classification of the branch of Articulata may therefore stand thus: —

1st Class: *Worms;* with three orders, Trematods (including Cestods, Planariæ, and Leeches), Nematoids (including Acanthocephala and Gordiacei), and Annelides.

2d Class: *Crustacea;* with four orders, Rotifera, Entomostraca (including Cirripeds), Tetradecapods, and Decapods.

3d Class: *Insects;* with three orders, Myriapods, Arachnids, and Insects proper.

There is not a dissenting voice among anatomists respecting the natural limits of the Vertebrata as a branch of the animal kingdom. Their character, however, does not so much consist in the structure of their backbone or the presence of a dorsal cord as in the general plan of that structure, which exhibits a cavity above and a cavity below a solid axis. These two cavities are circumscribed by complicated arches, arising from the axis, which are made up of different systems of organs, the skeleton, the muscles, vessels, and nerves, and include, the upper one the centres of the nervous system, the lower one the different systems of organs by which assimilation and reproduction are carried on.

The number and limits of the classes of this branch are not yet satisfactorily ascertained. At least naturalists do not all agree about them. For my part, I believe that the Marsupialia cannot be separated from the Placental Mammalia as a distinct class, since we observe, within the limits of another type of Vertebrata, the Selachians, which cannot be subdivided into classes, similar differences in the mode of development to those which exist between the Marsupials and the other Mammalia. But I hold at the same time with other naturalists that the Batrachia must be separated as a class from the true Reptiles, as the characters which distinguish them are of the kind upon which classes are founded. I am also satisfied that the differences which exist between the Selachians (the Skates, Sharks, and Chimæræ) are of the same kind as those which distinguish the Amphibians from the Reptiles proper, and justify therefore their separation as a class from the Fishes proper. I consider also the Cyclostomes as a distinct class for similar reasons; but I am still doubtful whether the Ganoids should be separated also from the ordinary Fishes. This, however, cannot be decided until their embryological development has been thoroughly investigated, though I have already collected data which favor this view of the case. Should this expectation be realized, the branch of Vertebrata would contain the following classes: —

1st Class: *Myzontes;* with two orders, Myxinoids and Cyclostomes.

2d Class: *Fishes* proper; with two orders, Ctenoids and Cycloids.[10]

3d Class: *Ganoids;* with three orders, Cœlacanths, Acipenseroids, and Sauroids; and doubtful, the Siluroids, Plectognaths, and Lophobranches.[11]

4th Class: *Selachians;* with three orders, Chimæræ, Galeodes, and Batides.

5th Class: *Amphibians;* with three orders, Cæciliæ, Ichthyodi, and Anura.

6th Class: *Reptiles;* with four orders, Serpentes, Saurii, Rhizodontes, and Testudinata.

7th Class: *Birds;* with four orders, Natatores, Grallæ, Rasores, and Insessores (including Scansores and Accipitres).

8th Class: *Mammalia;* with three orders, Marsupialia, Herbivora, and Carnivora.

I shall avail myself of an early opportunity to investigate more fully how far these groups of Vertebrata exhibit such characters as distinguish classes; and I submit my present impressions upon this

[10] I am satisfied that this subdivision of the Fishes proper requires modifications; but I fear it would lead me too far, were I to discuss here the reasons for the changes I propose to introduce into it.

[11] I have observed a very curious and peculiar mode of locomotion in all the Lophobranches, Scleroderms, and Gymnodontes which I have seen alive. They do not progress by the lateral motions of the vertebral columns, as other fishes do; but chiefly by an undulatory movement of their vertical fins, resembling very much the mode of action of the vibratile membranes. In this they resemble the young Lepidosteus; and I consider this fact as a new argument in favor of their association with the true Ganoids.

subject, rather as suggestions for further researches, than as matured results.[12]

SECTION II

EARLY ATTEMPTS TO CLASSIFY ANIMALS

So few American naturalists have paid special attention to the classification of the animal kingdom in general, that I deem it necessary to allude to the different principles which at different times have guided zoologists in their attempts to group animals according to their natural affinities. This will appear the more acceptable, I hope, since few of our libraries contain even the leading works of our science, and many zealous students are thus prevented from attempting to study what has thus far been done.

Science has begun in the introduction of names to designate natural groups of different value with the same vagueness which still prevails in ordinary language in the use of class, order, genus, family, species; taking them either as synonyms or substituting one for the other at random. Linnæus was the first to urge upon naturalists precision in the use of four kinds of groups in natural history, which he calls classes, orders, genera, and species.

Aristotle, and the ancient philosophers generally, distinguished only two kinds of groups among animals, *γένος* and *εδος* (genus and species). But the term genus had a most unequal meaning, applying at times indiscriminately to any extensive group of species and designating even what we now call classes as well as any other minor group. In the sense of class, it is taken in the following case: *λέγω δὲ γένος, οἷον ὄρνιθα, καὶ ἰχθῦν,*[13] while *εἶδος* is generally used for species, as the following sentence shows: *καὶ ἔστιν εἴδη πλείω ἰχθύων καὶ ὀρνίθων,*[14] though it has occasionally also a wider meaning. The sixth chapter of book I is the most important in the whole work of Aristotle

[12] [The preceding section represents Agassiz's effort to formulate a modern system of classification. It was based primarily on the taxonomic ideas and contributions of Cuvier, but contained corrections of that naturalist's work and material drawn from investigations of Agassiz and others made after Cuvier had published.]

[13] ["By 'genus' I mean, for instance, Bird or Fish." Thompson, tr., *Hist. Anim.*, I. 1. 486^a24.]

[14] [" . . . and there are many species of fishes and of birds." *Ibid.*, I. 1. 486^a24.]

upon this subject, as it shows to how many different kinds of groups the term *γένος* is applied. Here he distinguishes between *γένη μέγιστα* and *γένη μεγάλα* and *γένος* shortly. *Γένη δὲ μέγιστα τῶν ζώων, εἰς ἃ διαιρεῖται τἄλλα ζῶα, τάδ' ἐστίν· ἓν μὲν ὀρνίθων, ἓν δ' ἰχθύων, ἄλλο δὲ κήτους. ἄλλο δὲ γένος ἐστὶ τὸ τῶν ὀστρακοδέρμων. . . . Τῶν δὲ λοιπῶν ζώων οὐκ ἔστι τὰ γένη μεγάλα· οὐ γὰρ περιέχει πολλὰ εἴδη ἓν εἶδος, . . . τὰ δ' ἔχει μέν, ἀλλ' ἀνώνυμα.*[15] This is further insisted upon anew: *τοῦ δὲ γένους τῶν τετραπόδων ζώων καὶ ζωοτόκων εἴδη μὲν εἰσι πολλὰ, ἀνώνυμα δὲ.*[16] Here *εἶδος* has evidently a wider meaning than our term species, and the accurate Scaliger[17] translates it by *genus medium,* in contradistinction to *γένος*, which he renders by *genus summum. Εἶδος*, however, is generally used in the same sense as now, and Aristotle already considers fecundity as a specific character when he says of the Hemionos that it is called so from its likeness to the Ass, and not because it is of the same species; for, he adds, they copulate and propagate among themselves: *αἵ καλοῦνται ἡμίονοι δι' ὁμοιότητα, οὐκ οὖσαι ἁπλῶς τὸ αὐτὸ εἶδος· καὶ γὰρ ὀχεύονται καὶ γεννῶνται ἐξ ἀλλήλων.*[18] In another passage *γένος* applies, however, to a group exactly identical with our modern genus Equus: *ἐπεὶ ἐοτιν ἕν τι γένος καὶ ἐπὶ τοῖς ἔχουσι χαίτην, λοφούροις καλουμένοις, οἷον ἵππῳ καὶ ὄνῳ καὶ ὀρεῖ και γίννῳ καὶ ἴννῳ και τοῖς ἐν Συρίᾳ καλουμέναις ἡμιόνοις.*[19]

[15] ["Very extensive genera of animals, into which other subdivisions fall, are the following: one, of birds, one, of fishes, and another of Cetaceans. There is another genus of the hard-shell kind. . . . Of the other animals the genera are not extensive. For in them one species does not comprehend many species. . . . [but in one case, as man, the species is simple, admitting of no differentiation, while other cases admit of differentiation,] but the forms lack particular designations." *Ibid.,* I. 6. 490[b] Thompson points out that this entire passage is very troublesome, as Aristotle seems to juggle the terms *eidos* and *genos.*]

[16] ["In the genus that combines all viviparous quadrupeds are many species . . . but under no common appellation." *Ibid.,* I. 6. 490[b].]

[17] [Julius Caesar Scaliger, 1484–1558, tr., *Historia Animalium* (Toulouse, 1619), and *Histoire des Animaux* (Toulouse, 1619).]

[18] [" . . . from their externally resembling mules, though they are not strictly of the same species. And that they are not so is proved by the fact that they mate with and breed from one another." Thompson, *op. cit.,* I. 6. 491[a].]

[19] ["There is a sort of genus that embraces all creatures that have bushy manes and bushy tails, such as the horse, the ass, the mule, the jennet and the animals that are called Hemioni in Syria." Thompson, *loc. cit.* (Both Scaliger and Thompson point out that the purpose of Aristotle's discussion in this sixth chapter is not to define genera but rather to denominate them by showing how familiar animals correspond to real generic groups.)]

Aristotle cannot be said to have proposed any regular classification. He speaks constantly of more or less extensive groups, under a common appellation, evidently considering them as natural divisions; but he nowhere expresses a conviction that these groups may be arranged methodically so as to exhibit the natural affinities of animals. Yet he frequently introduces his remarks respecting different animals in such an order and in such connections as clearly to indicate that he knew their relations. When speaking of Fishes, for instance, he never includes the Selachians.

After Aristotle, the systematic classification of animals makes no progress for two thousand years, until Linnæus introduces new distinctions and assigns a more precise meaning to the terms class (*genus summum*), order (*genus intermedium*), genus (*genus proximum*), and species, the two first of which are introduced by him for the first time as distinct groups under these names in the system of Zoology.

SECTION III

PERIOD OF LINNÆUS

When looking over the *Systema Naturæ* of Linnæus, taking as the standard of our appreciation even the twelfth edition, which is the last he edited himself, it is hardly possible in our day to realize how great was the influence of that work upon the progress of Zoology.[20] And yet it acted like magic upon the age and stimulated exertions far surpassing anything that had been done in preceding centuries. Such a result must be ascribed partly to the circumstance that he was the first man who ever conceived distinctly the idea of expressing in a definite form what he considered to be a system of nature, and partly also to the great comprehensiveness, simplicity, and clearness of his method. Discarding in his system everything that could

[20] To appreciate correctly the successive improvements of the classification of Linnæus, we need only compare the first edition of the *Systema Naturæ* (1735), with the second (1740), the sixth (1748), the tenth (1758), and the twelfth (1766–1768), as they are the only editions he revised himself. The third is only a reprint of the first, the fourth and fifth are reprints of the second; the seventh, eighth, and ninth are reprints of the sixth; the eleventh is a reprint of the tenth; and the thirteenth, published after his death by Gmelin (1783–1793), is a mere compilation, deserving little confidence.

not easily be ascertained, he for the first time divided the animal kingdom into distinct classes, characterized by definite features; he also for the first time introduced orders into the system of Zoology besides genera and species, which had been vaguely distinguished before.[21] And though he did not even attempt to define the characteristics of these different kinds of groups, it is plain from his numerous writings that he considered them all as subdivisions of a successively more limited value, embracing a larger or smaller number of animals, agreeing in more or less comprehensive attributes. He expresses his views of these relations between classes, orders, genera, species, and varieties, by comparisons, in the following manner.[22]

Classis	*Ordo*	*Genus*	*Species*	*Varietas*
Genus summum	Genus intermedium	Genus proximum	Species	Individuum
Provinciæ	Territoria	Parœciæ	Pagi	Domicilium
Legiones	Cohortes	Manipuli	Contubernia	Miles

His arrangement of the animal kingdom is presented in the following diagram, compiled from the twelfth edition, published in 1766.

CLASSIFICATION OF LINNÆUS

Cl. 1. **Mammalia.** *Ord.* Primates, Bruta, Feræ, Glires, Pecora, Belluæ, Cete.
Cl. 2. **Aves.** *Ord.* Accipitres, Picæ, Anseres, Grallæ, Gallinæ, Passeres.
Cl. 3. **Amphibia.** *Ord.* Reptiles, Serpentes, Nantes.
Cl. 4. **Pisces.** *Ord.* Apodes, Jugulares, Thoracici, Abdominales.
Cl. 5. **Insecta.** *Ord.* Coleoptera, Hemiptera, Lepidoptera, Neuroptera, Hymenoptera, Diptera, Aptera.
Cl. 6. **Vermes.** *Ord.* Intestina, Mollusca, Testacea, Lithophyta, Zoophyta.

In the earlier editions, up to the tenth, the class of Mammalia was called Quadrupedia and did not contain the Cetaceans, which were still included among the Fishes. There seems never to have existed any discrepancy among naturalists respecting the natural limits of the class of Birds since it was first characterized by Linnæus, in a manner which excluded the Bats and referred them to the class of Mammalia. In the early editions of the *Systema Naturæ* the class of Reptiles embraces the same animals as in the systems of the most recent investigators; but since the tenth edition it has been encumbered with the addition of the cartilaginous and semicartilagi-

[21] The γένη μέγιστα of Aristotle correspond, however, to the classes of Linnæus; the γένη μεγάλα to his orders.
[22] See *Systema Naturæ* (12th ed.) p. 13.

nous Fishes, a retrograde movement suggested by some inaccurate observations of Dr. Garden.[23] The class of Fishes is very well limited in the early editions of the *Systema,* with the exception of the admission of the Cetaceans (Plagiuri) which were correctly referred to the class of Mammalia in the tenth edition. In the later editions, however, the Cyclostoms, Plagiostoms, Chimæræ, Sturgeons, Lophioids, Discoboli, Gymnodonts, Scleroderms, and Lophobranches are excluded from it and referred to the class of Reptiles. The class of Insects,[24] as limited by Linnæus, embraces not only what are now considered as Insects proper, but also the Myriapods, the Arachnids, and the Crustacea; it corresponds more accurately to the division of Arthropoda of modern systematists. The class of Worms, the most heterogeneous of all, includes besides all Radiata or Zoophytes and the Mollusks of modern writers, also the Worms, intestinal and free, the Cirripeds, and one Fish (Myxine). It was left for Cuvier[25] to introduce order in this chaos.

Such is, with its excellences and shortcomings, the classification which has given the most unexpected and unprecedented impulse to the study of Zoology. It is useful to remember how lately even so imperfect a performance could have so great an influence upon the progress of science, in order to understand why it is still possible that so much remains to be done in systematic Zoology. Nothing indeed can be more instructive to the student of Natural History than a careful and minute comparison of the different editions of the *Systema Naturæ* of Linnæus, and of the works of Cuvier and other prominent zoologists, in order to detect the methods by which real progress is made in our science.

Since the publication of the *Systema Naturæ* up to the time when Cuvier published the results of his anatomical investigations, all the attempts at new classifications were, after all, only modifications of

[23] [Alexander Garden, 1730?–1791.]

[24] Aristotle divides this group more correctly than Linnæus, as he admits already two classes among them, the Malacostraca (Crustacea) and the Entoma (Insects). *Hist. Anim.,* Lib. I, Chap. 6. VI. He seems also to have understood correctly the natural limits of the classes of Mammalia and Reptiles, for he distinguishes the Viviparous and Oviparous Quadrupeds and nowhere confounds Fishes with Reptiles. *Ibid.*

[25] It would be injustice to Aristotle not to mention that he understood already the relations of the animals united in one class by Linnæus, under the name of Worms, better than the great Swedish naturalist. Speaking, for instance, of the great genera or classes, he separates correctly the Cephalopods from the other Mollusks, under the name of Malakia. *Hist. Anim.,* Lib. I., Chap. 6.

the principles introduced by Linnæus in the systematic arrangement of animals. Even his opponents labored under the influence of his master spirit, and a critical comparison of the various systems which were proposed for the arrangement of single classes or of the whole animal kingdom shows that they were framed according to the same principles, namely, under the impression that animals were to be arranged together into classes, orders, genera, and species, according to their more or less close external resemblance. No sooner, however, had Cuvier presented to the scientific world his extensive researches into the internal structure of the whole animal kingdom than naturalists vied with one another in their attempts to remodel the whole classification of animals, establishing new classes, new orders, new genera, describing new species, and introducing all manner of intermediate divisions and subdivisions under the name of families, tribes, sections, etc. Foremost in these attempts was Cuvier himself, and next to him Lamarck. It has, however, often happened that the divisions introduced by the latter under new names were only translations into a more systematic form of the results Cuvier had himself obtained from his dissections and pointed out in his *Leçons sur l'anatomie comparée,* as natural divisions, but without giving them distinct names. Cuvier beautifully expresses the influence which his anatomical investigations had upon Zoology, and how the improvements in classification have contributed to advance comparative anatomy, when he says in the preface to the *Règne animal,*" page vi.: "Je dus donc, et cette obligation me prit un temps considérable, je dus faire marcher de front l'anatomie et la zoologie, les dissections et le classement; chercher dans mes premières remarques sur l'organisation, des distributions meilleures; m'en servir pour arriver à des remarques nouvelles; employer encore ces remarques à perfectionner les distributions; faire sortir enfin de cette fécondation mutuelle des deux sciences l'une par l'autre, un système zoologique propre à servir d'introducteur et de guide dans le champ de l'anatomie, et un corps de doctrine anatomique propre à servir de développement et d'explication au système zoologique." [26]

[26] ["I had then, and this obligation took me considerable time, I had to make anatomy and zoology, dissections and classification, move abreast of each other; to search in my first observations on organization for better arrangements through which to arrive at new observations, and to use again these observations to perfect the arrangements; finally, to take from this mutual fertilization between the two sciences, the one by

Without entering into a detailed account of all that was done in this period towards improving the system of Zoology, it may suffice to say that before the first decade of this century had passed, more than twice as many classes as Linnæus adopted had been characterized in this manner. These classes are: the Mollusks, Cirripeds, Crustacea, Arachnids, Annelids, Entozoa (Intestinal Worms), Zoophytes, Radiata, Polyps and Infusoria. Cuvier[27] admitted at first only eight classes, Duméril[28] nine, Lamarck[29] eleven and afterwards fourteen. The Cephalopoda, Gasteropoda, and Acephala, first so named by Cuvier, are in the beginning considered by him as orders only in the class of Mollusks; the Echinoderms also, though for the first time circumscribed by him within their natural limits, constitute only an order of the class of Zoophytes, not to speak of the lowest animals, which, from want of knowledge of their internal structure, still remain in great confusion. In this rapid sketch of the farther subdivisions which the classes Insecta and Worms of Linnæus have undergone under the influence of Cuvier, I have not, of course, alluded to the important contributions made to our knowledge of isolated classes by special writers, but limited my remarks to the works of those naturalists who have considered the subject upon the most extensive scale.

Thus far no attempt had been made to combine the classes among themselves into more comprehensive divisions, under a higher point of view, beyond that of dividing the whole animal kingdom into Vertebrata and Invertebrata, a division which corresponds to that of Aristotle, into ζῶα ἔναιμα and ζῶα ἄναιμα. All efforts were rather directed towards establishing a natural series, from the lowest Infusoria up to Man; which with many soon became a favorite tendency and ended by being presented as a scientific doctrine by Blainville.

the other, a zoological system appropriate as an introduction and guide in the field of anatomy, and a body of anatomical doctrine appropriate for the development and explication of the zoological system."]

[27] *Tableau élémentaire de l'histoire naturelle des animaux* (Paris, 1798).

[28] *Zoologie analytique* . . . (Paris, 1806).

[29] *Système des animaux sans vertèbres ou tableau général* . . . (Paris, 1801), and *Histoire naturelle des animaux sans vertèbres* . . . (7 vols., Paris, 1815–1822).

SECTION IV

PERIOD OF CUVIER AND ANATOMICAL SYSTEMS

The most important period in the history of Zoology begins, however, with the year 1812, when Cuvier laid before the Academy of Sciences in Paris the results of his investigations upon the more intimate relations of certain classes of the animal kingdom to one another,[30] which had satisfied him that all animals are constructed upon four different plans, or, as it were, cast in four different moulds. A more suggestive view of the subject never was presented before to the appreciation of investigators; and though it has by no means as yet produced all the results which certainly are to flow from its further consideration, it has already led to the most unquestionable improvements which classification in general has made since the days of Aristotle, and, if I am not greatly mistaken, it is only in as far as that fundamental principle has been adhered to that the changes proposed in our systems by later writers have proved a real progress, and not as many retrograde steps.

This great principle, introduced into our science by Cuvier, is expressed by him in these memorable words: "Si l'on considère le règne animal d'après les principes que nous venons de poser, en se débarrassant des préjugés établis sur les divisions anciennement admises, en n'ayant égard qu'a l'organisation et à la nature des animaux, et non pas à leur grandeur, à leur utilité, au plus ou moins de connaissance que nous en avons, ni à toutes les autres circonstances accessoires, on trouvera qu'il existe quatre formes principales, quatre plans généraux, si l'on peut s'exprimer ainsi, d'après lesquels tous les animaux semblent avoir été modelés et dont les divisions ultérieures, de quelque titre que les naturalistes les aient décorées, ne sont que des modifications assez légères fondées sur le développement ou l'addition de quelques parties, qui ne changent rien à l'essence du plan." [31]

[30] *Annales du muséum d'histoire naturelle,* XIX (1812), 73.

[31] ["If one considers the animal kingdom according to the principles that we are proposing, getting rid of the prejudices established by the divisions anciently admitted, in having only to consider the organization and nature of animals, and not their size, their use, whatever knowledge we have of them or all the other additional circum-

It is therefore incredible to me how, in presence of such explicit expressions, Cuvier can be represented, as he is still occasionally, as favoring a division of the animal kingdom into Vertebrata and Invertebrata.[32] Cuvier, moreover, was the first to recognize practically the inequality of all the divisions he adopts in his system; and this constitutes further a great and important step, even though he may not have found the correct measure for all his groups. For we must remember that at the time he wrote naturalists were bent upon establishing one continual uniform series to embrace all animals, between the links of which it was supposed there were no unequal intervals. The watchword of their school was: *Natura non facit saltum.* They called their system *la chaine des êtres.*

The views of Cuvier led him to the following arrangement of the animal kingdom: —

CLASSIFICATION OF CUVIER[33]

First Branch. ANIMALIA VERTEBRATA.

CL. 1. **Mammalia.** *Orders:* Bimana, Quadrumana, Carnivora, Marsupialia, Rodentia, Edentata, Pachydermata, Ruminantia, Cetacea.
CL. 2. **Birds.** *Ord.* Accipitres, Passeres, Scansores, Gallinæ, Grallæ, Palmipedes.
CL. 3. **Reptilia.** *Ord.* Chelonia, Sauria, Ophidia, Batrachia.
CL. 4. **Fishes.** 1*st Series:* **Fishes proper.** *Ord.* Acanthopterygii; — Abdominales, Subbrachii, Apodes; — Lophobranchii, Plectognathi; 2*d Series:* **Chondropterygii.** *Ord.* Sturiones, Selachii, Cyclostomi.[34]

Second Branch. ANIMALIA MOLLUSCA.

CL. 1. **Cephalopoda.** No subdivisions into orders or families.
CL. 2. **Pteropoda.** No subdivisions into orders or families.
CL. 3. **Gasteropoda.** *Ord.* Pulmonata, Nudibranchia, Inferobranchia, Tectibranchia, Heteropoda, Pectinibranchia, Tubulibranchia, Scutibranchia, Cyclobranchia.
CL. 4. **Acephala.** *Ord.* Testacea, Tunicata.
CL. 5. **Brachiopoda.** No subdivisions into orders or families.
CL. 6. **Cirrhopoda.** No subdivisions into orders or families.

---

stances, one finds that there exist four principal forms, four general plans, if one can express it thus, according to which all the animals seem to have been modeled, whose subsequent divisions, whatever names naturalists have bestowed upon them, are only modifications slight enough, founded on the development or the addition of several parts, which do not change the essence of the plan."]

[32] Ehrenberg, *Die Corralenthiere . . .*, p. 30.

[33] *Le Règne animal* (2d ed.). The classes of Crustacea, Arachnids, and Insects have been elaborated by Latreille. For the successive modifications the classification of Cuvier has undergone, compare his *Tableau élémentaire . . .* (1798), his paper in *Annales du muséum . . .* XIX (1812), 73, and the first edition of the *Règne animal* published in 1817.

[34] Compare *Règne animal* (2d ed.) II, 128, 383.

Third Branch. ANIMALIA ARTICULATA.

CL. 1. **Annelides.** *Ord.* Tubicolæ, Dorsibranchiæ, Abranchiæ.
CL. 2. **Crustacea.** 1*st Section:* **Malacostraca.** *Ord.* Decapoda, Stomapoda, Amphipoda, Læmodipoda, Isopoda. 2*d Section:* **Entomostraca.** *Ord.* Branchiopoda, Poecilopoda, Trilobitæ.
CL. 3. **Arachnides.** *Ord.* Pulmonariæ, Tracheariæ.
CL. 4. **Insects.** *Ord.* Myriapoda, Thysanura, Parasita, Suctoria, Coleoptera, Orthoptera, Hemiptera, Neuroptera, Hymenoptera, Lepidoptera, Rhipiptera, Diptera.

Fourth Branch. ANIMALIA RADIATA.

CL. 1. **Echinoderms.** *Ord.* Pedicellata, Apoda.
CL. 2. **Intestinal Worms.** *Ord.* Nematoidea (incl. Epizoa and Entozoa) Parenchymatosa.
CL. 3. **Acalephae.** *Ord.* Simplices, Hydrostaticæ.
CL. 4. **Polypi.** (Including Anthozoa, Hydroids, Bryozoa, Corallinæ, and Spongiæ). *Ord.* Carnosi, Gelatinosi, Polypiarii.
CL. 5. **Infusoria.** *Ord.* Rotifera and Homogenea (including Polygastrica and some Algæ.)

When we consider the zoological systems of the past century, that of Linnæus, for instance, and compare them with more recent ones, that of Cuvier, for example, we cannot overlook the fact that even when discoveries have added little to our knowledge, the subject is treated in a different manner; not merely in consequence of the more extensive information respecting the internal structure of animals, but also respecting the gradation of the higher groups.

Linnæus had no divisions of a higher order than classes. Cuvier introduced for the first time four great divisions, which he called *"embranchements"* or branches, under which he arranged his classes, of which he admitted three times as many as Linnæus had done.

Again, Linnæus divides his classes into orders; next, he introduces genera, and finally, species; and this he does systematically in the same gradation through all classes, so that each of his six classes is subdivided into orders, and these into genera with their species. Of families, as now understood, Linnæus knows nothing.

The classification of Cuvier presents no such regularity in its framework. In some classes he proceeds, immediately after presenting their characteristics, to the enumeration of the genera they contain, without grouping them either into orders or families. In other classes he admits orders under the head of the class, and then proceeds to the characteristics of the genera, while in others still, he admits under the class not only orders and families, placing always the family in a subordinate position to the order, but also a number of secondary divisions which he calls sections, divisions, tribes, etc., before he reaches the genera and species. With reference to the genera again, we find marked discrepancies in different classes. Sometimes a genus is to him an extensive group of species, widely differing one from the other, and of such genera he speaks as "grands genres"; others are limited in their extent, and contain homogeneous species without farther subdivisions, while still others are subdivided into what he calls sub-genera, and this is usually the case with his "great genera."

The gradation of divisions with Cuvier varies then with his classes, some classes containing only genera and species, and neither orders nor families nor any other subdivision. Others contain orders, families, and genera, and besides these, a variety of subdivisions of the most diversified extent and significance. This remarkable inequality between all the divisions of Cuvier is, no doubt, partly owing to the state of Zoology and of zoological museums at the time he wrote, and to his determination to admit into his work only such representatives of the animal kingdom as he could to a greater or less extent examine anatomically for himself; but it is also partly to be ascribed to his conviction, often expressed, that there is no such uniformity or regular serial gradation among animals as many naturalists attempted to introduce into their classifications.

CLASSIFICATION OF LAMARCK[35]

INVERTEBRATA.

I. Apathetic Animals.	
Cl. 1. **Infusoria.** *Ord.* Nuda, Appendiculata.	Do not feel, and move only by their excited irritability. No brain, nor elongated medullary mass; no senses; forms varied; rarely articulations.
Cl. 2. **Polypi.** *Ord.* Ciliati (Rotifera), Denudati (Hydroids), Vaginati (Anthozoa and Bryozoa), and Natantes (Crinoids, and some Halcyonoids.)	
Cl. 3. **Radiaria.** *Ord.* Mollia (Acalephæ), Echinoderms (including Holothuriæ and Actiniæ.)	
Cl. 4. **Tunicata.** *Ord.* Bothryllaria (Compound Ascidians), Ascidia (Simple Ascidians.)	
Cl. 5. **Vermes.** *Ord.* Molles and Rigiduli (Intestinal Worms and Gordius), Hispiduli (Nais), Epizoariæ (Epizoa, Lernæans.)	

[35] Lamarck, *Histoire naturelle des animaux sans vertèbres* . . . (7 vols., Paris, 1815–1822). A second edition with notes has been published by Gérard P. Deshayes and Milne-Edwards (10 vols., Paris, 1835–1843). For the successive modifications this classification has undergone see also Lamarck, *Systeme des animaux sans vertèbres* . . . , *Philosophie zoologique* . . . (2 vols., Paris, 1809), and *Extrait du Cours de Zoologie du muséum d'Histoire naturelle* . . . (Paris, 1812).

II. Sensitive Animals.

Cl. 6. **Insects.** (Hexapods.) *Ord.* Aptera, Diptera, Hemiptera, Lepidoptera, Hymenoptera, Neuroptera, Orthoptera, Coleoptera.
Cl. 7. **Arachnids.** *Ord.* Antennato-tracheales (Thysanura and Myriapoda), Exantennato-tracheales and Exantennato-branchiales (Arachnids proper.)
Cl. 8. **Crustacea.** *Ord.* Heterobranchia (Branchipoda, Isopoda, Amphipoda, Stomapoda) and Homobranchia (Decapoda.)
Cl. 9. **Annelids.** *Ord.* Apoda, Antennata, Sedentaria.
Cl. 10. **Cirripeds.** *Ord.* Sessilia and Pedunculata.
Cl. 11. **Conchifera.** *Ord.* Dimyaria, Monomyaria.
Cl. 12. **Mollusks.** *Ord.* Pteropoda, Gasteropoda, Trachelipoda, Cephalopoda, Heteropoda.

Feel, but obtain from their sensations only perceptions of objects, a sort of simple ideas, which they are unable to combine to obtain complex ones. No vertebral column; a brain and mostly an elongated medullary mass; some distinct senses; muscles attached under the skin; form symmetrical, the parts being in pairs.

VERTEBRATA.

III. Intelligent Animals.

Cl. 13. **Fishes.**
Cl. 14. **Reptiles.**
Cl. 15. **Birds.**
Cl. 16. **Mammalia.**

Feel; acquire preservable ideas; perform with them operations by which they obtain others; are intelligent in different degrees. A vertebral column; a brain and a spinal marrow; distinct senses; the muscles attached to the internal skeleton; form symmetrical, the parts being in pairs.

It is not easy to appreciate correctly the system of Lamarck, as it combines abstract conceptions with structural considerations and an artificial endeavor to arrange all animals in continuous series. The primary subdivision of the animal kingdom into Invertebrata and Vertebrata corresponds, as I have stated above, to that of *Anaima* and *Enaima* of Aristotle. The three leading groups designated under the name of Apathetic, Sensitive, and Intelligent animals, are an imitation of the four branches of Cuvier; but far from resting upon such a definite idea as the divisions of Cuvier, which involve a special plan of structure, they are founded upon the assumption that the psychical faculties of animals present a serial gradation, which, when applied as a principle of classification, is certainly not admissible. To say that neither Infusoria, nor Polypi, nor Radiata, nor Tunicata, nor Worms feel is certainly a very erroneous assertion. They manifest sensations quite as distinctly as many of the animals included in the second type which are called Sensitive. And as to the other assertion, that they move only by their excited irritability, we need only watch the Starfishes to be satisfied that their motions are determined by internal impulses and not by external excitation. Modern investigations have shown that most of them have a nervous system, and many even organs of senses.

The Sensitive animals are distinguished from the third type, the Intelligent animals, by the character of their sensations. It is stated, in respect to the Sensitive animals, that they obtain from their sensations only perceptions of objects, a sort of simple ideas which they are unable to combine so as to derive from them complex ones, while the Intelligent animals are said to obtain ideas which they may preserve, and to perform with them operations by which they arrive at new ideas. They are said to be Intelligent. Even now, fifty years after Lamarck made those assertions, I doubt whether it is possible to distinguish in that way between the sensations of the Fishes, for instance, and those of the Cephalopods. It is true the structure of the animals called Sensitive and Intelligent by Lamarck differs greatly, but a large number of his Sensitive animals are constructed upon the same plan as many of those he includes among the Apathetic. They embrace, moreover, two different plans of structure, and animal psychology is certainly not so far advanced as to afford the least foundation for the distinctions here introduced.

Even from his own point of view, his arrangement of the classes is less perfect than he might have made it, as the Annelids stand nearer to the Worms than the Insects and are very inferior to them. Having failed to perceive the value of the idea of plan, and having substituted for it that of a more or less complicated structure, Lamarck unites among his Apathetic animals, Radiates (the Polypi and Radiaria) with Mollusks (the Tunicata) and with Articulates (the Worms). Among the Sensitive animals he unites Articulates (the Insects, Arachnids, Crustacea, Annelids, and Cirripeds) with Mollusks the Conchifera, and the Mollusks proper). Among the Intelligent animals he includes the ancient four classes of Vertebrates, the Fishes, Reptiles, Birds, and Mammalia.

CLASSIFICATION OF DE BLAINVILLE[36]

1. *Sub-Kingdom. Artiomorpha or Artiozoaria.* Form bilateral.

First Type: **Osteozoaria.** (Vertebrata.)
Sub-Type: *Vivipara.*
Cl. 1. **Pilifera,** or Mammifera. 1st. Monadelphya. 2d. Didelphya.
Sub-Type: *Ovipara.*
Cl. 2. **Pennifera,** or Aves.
Cl. 3. **Squamifera,** or Reptilia.
Cl. 4. **Nudipellifera,** or Amphibia.

[36] *De l'organisation des animaux* (1822).

CL. 5. **Pinnifera,** or Pisces.
Anosteozoaria.
Second Type: ENTOMOZOARIA. (Articulata.)
CL. 6. **Hexapoda.** (Insecta proprie sic dicta.)
CL. 7. **Octopoda.** (Arachnida.)
CL. 8. **Decapoda.** (Crustacea, Decapoda, and Limulus.)
CL. 9. **Heteropoda.** (Squilla, Entomostraca, and Epizoa.)
CL. 10. **Tetradecapoda.** (Amphipoda and Isopoda.)
CL. 11. **Myriapoda.**
CL. 12. **Chætopoda.** (Annelides.)
CL. 13. **Apoda.** (Hirudo, Cestoidea, Ascaris.)
Third Type: MALENTOZOARIA.
CL. 14. **Nematopoda.** (Cirripedia.)
CL. 15. **Polyplaxiphora.** (Chiton.)
Fourth Type: MALACOZOARIA. (Mollusca.)
CL. 16. **Cephalophora.** Dioïca, (Cephalopoda and Gasteropoda, p. p.) Hermaphrodita and Monoïca (Gasteropoda reliqua.)
CL. 17. **Acephalophora.** Palliobranchia (Brachiopoda), Lamellibranchia (Acephala), Heterobranchia (Ascidiæ.)

2. *Sub-Kingdom. Actinomorpha* or *Actinozoaria.* Form radiate.
CL. 18. **Annelidaria,** or Gastrophysaria (Sipunculus, etc.)
CL. 19. **Ceratodermaria.** (Echinodermata.)
CL. 20. **Arachnodermaria.** (Acalephæ.)
CL. 21. **Zoantharia.** (Actiniæ.)
CL. 22. **Polypiaria.** (Polypi tentaculis simplicibus), (Anthozoa and Bryozoa.)
CL. 23. **Zoophytaria.** (Polypi tentaculis compositis), (Halcyonoidea.)

3. *Sub-Kingdom. Heteromorpha* or *Heterozoaria.* Form irregular.
CL. 24. **Spongiaria.** (Spongiæ.)
CL. 25. **Monadaria.** (Infusoria.)
CL. 26. **Dendrolitharia.** (Corallinæ.)

The classification of de Blainville resembles those of Lamarck and Cuvier much more than a diagram of the three would lead us to suppose. The first of these systems is founded upon the idea that the animal kingdom forms one graduated series; only that de Blainville inverts the order of Lamarck, beginning with the highest animals and ending with the lowest. With that idea is blended, to some extent, the view of Cuvier that animals are framed upon different plans of structure but so imperfectly has this view taken hold of de Blainville that, instead of recognizing at the outset these great plans, he allows the external form to be the leading idea upon which his primary divisions are founded, and thus he divides the animal kingdom into three sub-kingdoms: the first, including his Artiozoaria, with a bilateral form; the second, his Actinozoaria, with a radiated form, and the third, his Heterozoaria, with an irregular form (the Sponges, Infusoria, and Corallines). The plan of structure is only introduced as a secondary consideration, upon which he establishes four types among the Artiozoaria: 1st. The Osteozoaria, corresponding to Cuvier's Vertebrata; 2d. The Entomozoaria, corresponding to Cuvier's Articulata; 3d. The Malentozoaria, which are a very artificial group, suggested only by the necessity of establishing a transition between the Articulata and Mollusca; 4th. The Malacozoaria, corresponding to Cuvier's Mollusca. The second sub-kingdom, Actinozoaria, corresponds to Cuvier's Radiata, while the third sub-kingdom, Heterozoaria, contains organized beings which for the most part do not belong to the animal kingdom. Such at least are his Spongiaria and Dendrolitharia, whilst his Monodaria answer to the old class of Infusoria, about which enough has already been said above. It is evident that what is correct in this general arrangement is borrowed from Cuvier; but it is only justice to de Blainville to say that in the limitation and arrangement of the classes he has introduced some valuable improvements. Among Vertebrata, for instance, he has for the first time distinguished the class of Amphibia from the true Reptiles. He was also the first to remove the Intestinal Worms from among the Radiata to the Articulata; but the establishment of a distinct type for the Cirripedia and Chitons was a very mistaken conception. Notwithstanding some structural peculiarities, the Chitons are built essentially upon the same plan as the Mollusks of the class Gasteropoda, and the investigations made not long after the publication of de Blainville's system have left no doubt that Cirripedia are genuine Crustacea. The supposed transition between Articulata and Mollusks, which de Blainville attempted to establish with his type of Malentozoaria, certainly does not exist in nature.

If we apply to the classes of de Blainville the test introduced in the preceding chapter, it will be obvious that his Decapoda, Heteropoda, and Tetradecapoda partake more of the character of orders than of that of classes, whilst among Mollusks his class Cephalophora certainly includes two classes, as he has himself acknowledged in his later works. Among Radiata his classes Żoantharia, Polypiaria, and Zoophytaria partake again of the character of orders and not of those of classes. One great objection to the system of de Blainville is the useless introduction of so many new names for groups which had already been correctly limited and well named by his predecessors. He had, no doubt, a desirable object in view in doing this — he wished to remove some incorrect names; but he extended his reform too far when he undertook to change those also which did not suit his system.

CLASSIFICATION OF EHRENBERG

The characteristics of the following twenty-eight classes of animals, with a twenty-ninth for Man alone, are given more fully in the *Transactions* of the Academy of Berlin for 1836.[37]

1st Cycle: NATIONS. Mankind, constituting one distinct class, is characterized by the equable development of all systems of organs, in contradistinction of the

2d Cycle: ANIMALS, which are considered as characterized by the prominence of single systems. These are divided into:

A. Myeloneura.

I. NUTRIENTIA. Warm-blooded Vertebrata, taking care of their young.
- CL. 1. **Mammalia.**
- CL. 2. **Birds.**

II. ORPHANOZOA. Cold-blooded Vertebrata, taking no care of their young.
- CL. 3. **Amphibia.**
- CL. 4. **Pisces.**

B. Ganglioneura.

A. Sphygmozoa, Cordata.

Circulation marked by a heart or pulsating vessels.

III. ARTICULATA. Real articulation, marked by rows of ganglia and their ramifications.
- CL. 5. **Insecta.**
- CL. 6. **Arachnoidea.**
- CL. 7. **Crustacea** (including Entomostraca, Cirripedia, and Lernæa.)
- CL. 8. **Annulata.** (The genuine Annelids exclusive of Nais.)
- CL. 9. **Somatotoma.** (Naidina.)

IV. MOLLUSCA. No articulation. Ganglia dispersed.
- CL. 10. **Cephalopoda.**
- CL. 11. **Pteropoda.**
- CL. 12. **Gasteropoda.**
- CL. 13. **Acephala.**
- CL. 14. **Brachiopoda.**
- CL. 15. **Tunicata.** (Ascidiæ simplices.)
- CL. 16. **Aggregata.** (Ascidiæ compositæ.)

B. Asphyeta, Vasculosa.

Vessels without pulsation.

V. TUBULATA. No real articulation. Intestine, a simple sac or tube.
- CL. 17. **Bryozoa.**
- CL. 18. **Dimorph a.** (Hydroids.)
- CL. 19. **Turbellaria.** (Rhabdocœla: Derostoma, Turbella, Vortex.)
- CL. 20. **Nematoidea.** (Entozoa, with simple intestine; also Gordius and Anguillula.)
- CL. 21. **Rotatoria.**
- CL. 22. **Echinoidea.** (Echinus, Holothuria, Sipunculus.)

VI. RACEMIFERA. Intestine divided, or forked, radiating, dendritic, or racemose.
- CL. 23. **Asteroidea.**

[37] [In XIV, 605. See also *Das Naturreich des Menschen* (1835).]

CL. 24. **Acalephæ.**
CL. 25. **Anthozoa.**
CL. 26. **Trematodea.** (Entozoa with ramified intestine, also Cercaria.)
CL. 27. **Complanata.** (Dendrocœla: Planaria, etc.)
CL. 28. **Polygastrica.**

The system of Zoology published by Ehrenberg in 1836 presents many new views in almost all its peculiarities. The most striking of its features is the principle laid down that the type of development of animals is one and the same from Man to the Monad, implying a complete negation of the principle advocated by Cuvier, that the four primary divisions of the animal kingdom are characterized by different plans of structure. It is very natural that Ehrenberg, after having illustrated so fully and so beautifully as he did the natural history of so many organized beings, which up to the publication of his investigations were generally considered as entirely homogeneous, after having shown how highly organized and complicated the internal structure of many of them is, after having proved the fallacy of the prevailing opinions respecting their origin, should have been led to the conviction that there is, after all, no essential difference between these animals, which were then regarded as the lowest, and those which were placed at the head of the animal creation. The investigator who had just revealed to the astonished scientific world the complicated systems of organs which can be traced in the body of microscopically small Rotifera must have been led irresistibly to the conclusion that all animals are equally perfect, and have assumed, as a natural consequence of the evidence he had obtained, that they stand on the same level with one another, as far as the complication of their structure is concerned. Yet the diagram of his own system shows that he himself could not resist the internal evidence of their unequal structural endowment. Like all other naturalists, he places Mankind at one end of the animal kingdom and such types as have always been considered as low at the other end.

Man constitutes, in his opinion, an independent cycle, that of nations, in contradistinction to the cycle of animals, which he divides into MYELONEURA, those with nervous marrow (the Vertebrata) and GANGLIONEURA, those with ganglia (the Invertebrata). The Vertebrata he subdivides into *Nutrientia,* those which take care of their young, and *Orphanozoa,* those which take no care of their young, though this is not strictly true, as there are many Fishes and Reptiles which provide as carefully for their young as some of the Birds and Mammalia, though they do it in another way. The Invertebrata are subdivided into *Sphygmozoa,* those which have a heart or pulsating vessels, and *Asphycta,* those in which the vessels do not pulsate. These two sections are further subdivided: the first, into Articulata with real articulations and rows of ganglia, and Mollusks without articulation and with dispersed ganglia; the second, into Tubulata with a simple intestine, and Racemifera with a branching intestine. These characters, which Ehrenberg assigns to his leading divisions, imply necessarily the admission of a gradation among animals. He thus negatives, in the form in which he expresses the results of his investigations, the very principle he intends to illustrate by his diagram. The peculiar view of Ehrenberg, that all animals are equal in the perfection of their organization, might be justified, if it was qualified so as to imply a relative perfection, adapted in all to the end of their special mode of existence. As no one observer has contributed more extensively than Ehrenberg to make known the complicated structure of a host of living beings, which before him were almost universally believed to consist of a simple mass of homogeneous jelly, such a view would naturally be expected of him. But this qualified perfection is not what he means. He does not wish to convey the idea that all animals are equally perfect in their way, for he states distinctly that "Infusoria have the same sum of systems of organs as Man," and the whole of his system is intended to impress emphatically this view. The separation of Man from the animals, not merely as a class but as a still higher division, is especially maintained upon that ground.

The principle of classification adopted by Ehrenberg is purely anatomical; the idea of type is entirely set aside, as is shown by the respective position of his classes. The Myeloneura, it is true, correspond to the branch of Vertebrata, and the Sphygmozoa to the Articulata and Mollusca; but they are not brought together on the ground of the typical plan of their structure, but because the first have a spinal marrow and the other a heart or pulsating vessels with or without articulations of the body. In the division of Tubulata it is still more evident how the plan of their structure is disregarded, as that section embraces Radiata (the Echinoidea and the Dimorphæa) Mollusca (the Bryozoa), and Articulata (the Turbellaria, the Nematoidea, and the Rotatoria), which are thus combined simply on the ground that they have vessels which do not pulsate and that their intestine is a simple sac or tube. The Racemifera contain also animals constructed upon different plans, united on account of the peculiar structure of the intestine, which is either forked or radiating, dendritic or racemose.

The limitation of many of the classes proposed by Ehrenberg is quite objectionable when tested by the principles discussed above. A large proportion of them are indeed founded upon ordinal characters only, and not upon class characters. This is particularly evident with the Rotatoria, the Somatotoma, the Turbellaria, the Nematoidea, the Trematodea, and the Complanata, all of which belong to the branch of Articulata. The Tunicata, the Aggregata, the Brachiopoda, and the Bryozoa are also only orders of the class Acephala. Before Echinoderms had been so extensively studied as of late, the separation of the Echinoidea from Asteroidea might have seemed justifiable; at the present day it is totally inadmissible. Even Leuckart, who considers the Echinoderms as a distinct branch of the animal kingdom, insists upon the necessity of uniting them as a natural group. As to the Dimorphæa, they constitute a natural order of the class Acalephæ, which is generally known by the name of Hydroids.

CLASSIFICATION OF BURMEISTER

The following diagram is compiled from the author's *Geschichte der Schöpfung* (Leipzig, 1843).

Type I. Irregular Animals.

1st Subtype. Cl. 1. **Infusoria.**

Type II. Regular Animals.

2d Subtype. Cl. 2. **Polypina.** *Ord.* Bryozoa, Anthozoa.
3d Subtype. Cl. 3. **Radiata.** *Ord.* Acalephæ, Echinodermata, Scytodermata.

Type III. Symmetrical Animals.

4th Subtype. Cl. 4. **Mollusca.** *Ord.* Perigymna (Tunicata); Cormopoda (Acephala); Brachiopoda, Cephalophora (Pteropoda and Gasteropoda); Cephalopoda.
5th Subtype. **Arthrozoa.**
Cl. 5. **Vermes.** *Ord.* Helminthes, Trematodes, and Annulati.
Cl. 6. **Crustacea.** 1°. **Ostracoderma.** *Ord.* Prothesmia (Cirripedia, Siphonostoma, and Rotatoria); Aspidostraca (Entomostraca: Lophyropoda, Phyllopoda, Pæcilopoda, Trilobitæ.) 2°. **Malacostraca.** *Ord.* Thoracostraca (Podophthalma); and Arthrostraca (Edriophthalma.)
Cl. 7. **Arachnoda.** *Ord.* Myriapoda, Arachnidæ.
Cl. 8. **Insecta.** *Ord.* Rhynchota, Synistata, Antliata, Piezata, Glossata, Eleutherata.
6th Subtype. **Osteozoa.** (Vertebrata.)

CL. 9. **Pisces.**
CL. 10. **Amphibia.**
CL. 11. **Aves.**
CL. 12. **Mammalia.**

The general arrangement of the classification of Burmeister recalls that of de Blainville; only that the order is inverted. His three types correspond to the three subkingdoms of de Blainville: the Irregular Animals to the Heterozoaria, the Regular Animals to the Actinozoaria, and the Symmetrical Animals to the Artiozoaria; while his subtypes of the Symmetrical Animals correspond to the types de Blainville admits among his Artiozoaria, with this important improvement, however, that the Malentozoaria are suppressed. Burmeister reduces, unhappily, the whole branch of Mollusks to one single class. The Arthrozoa, on the contrary, in the investigation of which Burmeister has rendered eminent service to science are presented in their true light. In his special works,[38] his classification of the Articulata is presented with more details. I have no doubt that the correct views he entertains respecting the standing of the Worms in the branch of Articulata are owing to his extensive acquaintance with the Crustacea and Insects and their metamorphoses.

CLASSIFICATION OF OWEN

The following diagram is compiled from R. OWEN'S *Lectures on the Comparative Anatomy and Physiology of the Invertebrate Animals* (2d edit., London, 1855).

Province. VERTEBRATA. Myelencephala. (Owen.)

CL. **Mammalia.**
CL. **Aves.**
CL. **Reptilia.**
} The classes Mammalia, Aves, and Reptilia are not yet included in the second volume of the "Lectures," the only one relating to Vertebrata thus far published.

CL. **Pisces.** *Ord.* Dermopteri, Malacopteri, Pharyngognathi, Anacanthini, Acanthopteri, Plectognathi, Lophobranchii, Ganoidei, Protopteri, Holocephali, Plagiostomi.

Province. ARTICULATA. Homogangliata. (Owen.)

CL. **Arachnida.** *Ord.* Dermophysa, Trachearia, Pulmotrachearia, and Pulmonaria.
CL. **Insecta.** *Subclass:* **Myriapoda.** *Ord.* Chilognatha and Chilopoda. *Subclass:* **Hexapoda.** *Ord.* Aptera, Diptera, Lepidoptera, Hymenoptera, Homoptera, Strepsiptera, Neuroptera, Orthoptera, and Coleoptera.
CL. **Crustacea.** *Subclass:* **Entomostraca.** *Ord.* Trilobites, Xiphosura, Phyllopoda, Cladocera, Ostracopoda, Copepoda. *Subclass:* **Malacostraca.** 1°. Edriophthalma. *Ord.* Læmodipoda, Isopoda, Amphipoda. 2°. Podophthalma. *Ord.* Stomapoda, Decapoda.
CL. **Epizoa.** *Ord.* Cephaluna, Brachiuna, and Onchuna.

[38] These are: *Beiträge zur Naturgeschichte . . .* (1834), *Handbuch der Entomologie* (5 vols., Berlin, 1832–1844), *Die Organisation der Trilobiten, aus ihren lebenden Verwandten entwickelt* (Berlin, 1843; English ed., London, 1847).

CL. **Annellata.** *Ord.* Suctoria, Terricola, Errantia, Tubicola.
CL. **Cirripedia.** *Ord.* Thoracica, Abdominalia, and Apoda.

Province. MOLLUSCA. Heterogangliata. (Owen.)

CL. **Cephalopoda.** *Ord.* Tetrabranchiata and Dibranchiata.
CL. **Gasteropoda.** A. **Monœcia:** *Ord.* Apneusta (Köll.), Nudibranchiata, Inferobranchiata, Tectibranchiata, Pulmonata. B. **Diœcia.** *Ord.* Nucleobranchiata, Tubulibranchiata, Cyclobranchiata, Scutibranchiata, and Pectinibranchiata.
CL. **Pteropoda.** *Ord.* Thecosomata and Gymnosomata.
CL. **Lamellibranchiata.** *Ord.* Monomyaria and Dimyaria.
CL. **Brachiopoda.** Only subdivided into families.
CL. **Tunicata.** *Ord.* Saccobranchiata and Tæniobranchiata.

Subprovince. RADIARIA.[39]

CL. **Echinodermata.** *Ord.* Crinoidea, Asteroidea, Echinoidea, Holothurioidea, and Sipunculoidea.
CL. **Bryozoa.** Only subdivided into families.
CL. **Anthozoa.** Only subdivided into families.
CL. **Acalephæ.** *Ord.* Pulmograda, Ciliograda, and Physograda.
CL. **Hydrozoa.** Only subdivided into families.

Subprovince. ENTOZOA.

CL. **Cœlelmintha.** *Ord.* Gordiacea, Nematoidea, and Onchophora.
CL. **Sterelmintha.** *Ord.* Tænioidea, Trematoda, Acanthocephala. — Turbellaria.

Subprovince. INFUSORIA.

CL. **Rotifera.** Only subdivided into families.
CL. **Polygastria.** *Ord.* Astoma, Stomatoda. — Rhizopoda.

The classification with which Owen[40] introduces his *Lectures on Comparative Anatomy* is very instructive, as showing, more distinctly than other modern systems, the unfortunate ascendancy which the consideration of the complication of structure has gained of late over the idea of plan. His provinces, it is true, correspond in the main to the branches of Cuvier, with this marked difference, however, that he does not recognize a distinct province of Radiata coequal with those of Mollusca, Articulata, and Vertebrata, but only admits Radiaria as a subprovince on a level with Entozoa and Infusoria. Here

[39] In the first edition of the work quoted above, published in 1843, the three subprovinces, Radiaria, Entozoa, and Infusoria are considered as one subkingdom called Radiata, in contradistinction of the subkingdoms, Mollusca, Articulata, and Vertebrata, and that subkingdom is subdivided into two groups, *Nematoneura* and *Acrita*.

[40] I have given precedence to the classification of Owen over those of von Siebold and Stannius, Milne-Edwards, Leuckart, etc., because the first edition of the *Lectures on Comparative Anatomy* was published in 1843; but in estimating its features, as expressed in the preceding diagram, it should be borne in mind that in the first edition the classes alone are considered, and that the orders and families were only added to the second edition in 1855. I mention this simply to prevent the possibility of being understood as ascribing to Owen all those subdivisions of the classes which he admits and which do not appear in the systems considered before his.

the idea of simplicity of structure evidently prevails over that of plan, as the subprovinces Radiaria, Entozoa, and Infusoria embrace, besides true Radiata, the lowest types of two other branches, Mollusks and Articulates. On the other hand, his three subprovinces correspond to the first three types of von Siebold; the Infusoria[41] of Owen embracing the same animals as the Protozoa of Siebold, his Entozoa[42] the same as the Vermes, and his Radiaria the same as the Zoophyta, with the single exception that Owen refers the Annellata to the province of Articulata, whilst Siebold includes them among his Vermes. Beyond this the types of Mollusca and Articulata (Arthropoda) of the two distinguished anatomists entirely agree. The position assigned by Owen to the provinces Articulata and Mollusca, not one above the other, but side by side with one another,[43] is no doubt meant to express his conviction, that the complication of structure of these two types does not justify the idea that either of them stands higher or lower than the other; and this is perfectly correct.

Several groups, established by previous writers as families or orders, are here admitted as classes. His class EPIZOA, which is not to be confounded with that established by Nitzsch under the same name, corresponds exactly to the family called LERNÉES by Cuvier. His class HYDROZOA answers to the order HYDROIA of Johnston and is identical with the class called DIMORPHÆA by Ehrenberg. His class CŒLELMINTHA corresponds to the order of INTESTINAUX CAVITAIRES established by Cuvier, with the addition of Gordius; while his class STERELMINTHA has the same circumscription as the order INTESTINAUX PARENCHYMATEUX of Cuvier. Generally speaking, it should not be understood that the secondary divisions mentioned by the different authors, whose systems I have analyzed here, were established by them. They are frequently borrowed from the results obtained by special investigators of isolated classes. But it would lead me too far to enter here into a discussion of all these details.

This growing resemblance of the modern systems of Zoology is a

[41] The Rhizopoda are considered as a group coequal to Rotifera and Polygastria, on p. 16 of the *Lectures* (1855) but on p. 59, they stand as a suborder of Polygastria.

[42] The Turbellaria are represented as an independent group, in *ibid.*, p. 16, and referred as a suborder to the Trematoda, on p. 118.

[43] From want of space, I have been compelled, in reproducing the classification of Owen in the preceding diagram, to place his provinces Articulata and Mollusca one below the other on my page; according to his views, they should stand on a level, side by side with one another.

very favorable sign of our times. It would indeed be a great mistake to assume that it is solely owing to the influence of different authors upon one another; it is, on the contrary, to a very great extent the result of our better acquaintance with Nature. When investigators at all conversant with the present state of our science must possess nearly the same amount of knowledge, it is self-evident that their views can no longer differ so widely as they did when each was familiar only with a part of the subject. A deeper insight into the animal kingdom must in the end lead to the conviction that it is not the task of zoologists to introduce order among animals, but that their highest aim should be simply to read the natural affinities which exist among them, so that the more nearly our knowledge embraces the whole field of investigation, the more closely will our opinions coincide.

As to the value of the classes adopted by Owen, I may further remark that recent investigations, of which he might have availed himself, have shown that the Cirripedia and his Epizoa are genuine Crustacea and that the Entozoa can no longer be so widely separated from the Annellata as in his system. With reference to the other classes, I refer the reader to my criticism of older systems and to the first section of this Chapter.

It is a great satisfaction for me to find that the views I have advocated in the preceding sections, respecting the natural relations of the leading groups of the animal kingdom, coincide so closely with the classification of that distinguished zoologist, Milne-Edwards, lately presented by him as the expression of his present views of the natural affinities of animals He is the only original investigator who has recently given his unqualified approbation to the primary divisions first proposed by Cuvier, admitting of course the rectifications among the group of secondary rank rendered necessary by the progress of science, to which he has himself so largely contributed.

As to the classes adopted by Milne-Edwards, I have little to add to what I have already stated before with reference to other classifications. Though no longer overruling the idea of plan, that of complication of structure has still too much influence with Milne-Edwards, inasmuch as it leads him to consider as classes groups of animals which differ only in degree and are therefore only orders. Such are no doubt his classes of Molluscoids and those of Worms, besides the Myriapods and Arachnids. Respecting the Fishes, I refer to my remarks in the first section of this Chapter.

CLASSIFICATION OF MILNE-EDWARDS

The following diagram is drawn from the author's *Cours élémentaire d'Histoire naturelle* (7th ed., Paris, 1855), in which he has presented the results of his latest investigations upon the classification of the Vertebrata and Articulata; the minor subdivisions of the Worms, Mollusks, and Zoophytes, however, are not considered in this work.

I. Osteozoaria, or Vertebrata.

Subbranch. ***Allantoidians.***

Cl. **Mammalia.** 1°. Monodelphya. *a.* Propria. *Ord.* Bimana, Quadrumana, Cheiroptera, Insectivora, Rodentia, Edentata, Carnivora, Amphibia, Pachydermata, Ruminantia. *b.* Pisciformia. *Ord.* Cetacea. 2°. Didelphya. *Ord.* Marsupialia, Monotremata.
Cl. **Birds.** *Ord.* Rapaces, Passeres, Scansores, Gallinæ, Grallæ, and Palmipedes.
Cl. **Reptiles.** *Ord.* Chelonia, Sauria, Ophidia.

Subbranch. ***Anallantoidians.***

Cl. **Batrachians.** *Ord.* Anura, Urodela, Perennibranchia, Cæciliæ.
Cl. **Fishes.** 1°. Ossei. *Ord.* Acanthopterygii, Abdominales, Subbrachii, Apodes, Lophobranchii, and Plectognathi. 2°. Chondropterygii. *Ord.* Sturiones, Selachii, and Cyclostomi.

II. Entomozoa, or Annellata.

Subbranch. ***Arthropoda.***

Cl. **Insecta.** *Ord.* Coleoptera, Orthoptera, Nevroptera, Hymenoptera, Lepidoptera, Hemiptera, Diptera, Rhipiptera, Anoplura, and Thysanura.
Cl. **Myriapoda.** *Ord.* Chilognatha and Chilopoda.
Cl. **Arachnids.** *Ord.* Pulmonaria and Trachearia.
Cl. **Crustacea.** 1°. Podophthalmia. *Ord.* Decapoda and Stomapoda. 2°. Edriophthalma. *Ord.* Amphipoda, Læmodipoda, and Isopoda. 3°. Branchiopoda. *Ord.* Ostrapoda, Phyllopoda, and Trilobitæ. 4°. Entomostraca. *Ord.* Copepoda, Cladocera, Siphonostoma, Lernæida, Cirripedia. 5°. Xiphosura.

Subbranch. ***Vermes.***

Cl. **Annelids.**
Cl. **Helminths.**
Cl. **Turbellaria.**
Cl. **Cestoidea.**
Cl. **Rotatoria.**

III. Malacozoaria, or Mollusca.

Subbranch. ***Mollusks proper.***

Cl. **Cephalopods.**
Cl. **Pteropods.**
Cl. **Gasteropods.**
Cl. **Acephala.**

Subbranch. ***Molluscoids.***

Cl. **Tunicata.**
Cl. **Bryozoa.**

IV. Zoophytes.

Subbranch. ***Radiaria,*** or ***Radiata.***

Cl. **Echinoderms.**
Cl. **Acalephs.**
Cl. **Corallaria,** or **Polypi.**

Subbranch. ***Sarcodaria.***

Cl. **Infusoria.**
Cl. **Spongiaria.**

CLASSIFICATION OF VON SIEBOLD AND STANNIUS

This classification is adopted in the following work: Siebold and Stannius, *Lehrbuch der vergleichenden Anatomie* (2 vols., Berlin, 1845; 2d ed., 1855).

EVERTEBRATA.

I. Protozoa.

Cl. 1. **Infusoria.** *Ord.* Astoma and Stomatoda.
Cl. 2. **Rhizopoda.** *Ord.* Monosomatia and Polysomatia.

II. Zoophyta.

Cl. 3. **Polypi.** *Ord.* Anthozoa and Bryozoa.
Cl. 4. **Acalephæ.** *Ord.* Siphonophora, Discophora, Ctenophora.
Cl. 5. **Echinodermata.** *Ord.* Crinoidea, Asteroidea, Echinoidea, Holothurioidea, and Sipunculoidea.

III. Vermes.

Cl. 6. **Helminthes.** *Ord.* Cystici, Cestodes, Trematodes, Acanthocephali, Gordiacei, Nematodes. — Since the publication of the work quoted above, Siebold has introduced most important improvements in the classification of the Worms, and greatly increased our knowledge of these animals.
Cl. 7. **Turbellarii.** *Ord.* Rhabdocœli, Dendrocœli.
Cl. 8. **Rotatorii.** Not subdivided into orders.
Cl. 9. **Annulati.** *Ord.* Apodes and Chætopodes.

IV. Mollusca.

Cl. 10. **Acephala.** *Ord.* Tunicata, Brachiopoda, Lamellibranchia.
Cl. 11. **Cephalophora,** Meck. (Gasteropoda.) *Ord.* Pteropoda, Heteropoda, Gasteropoda.
Cl. 12. **Cephalopoda.** Not subdivided into orders.

V. Arthropoda.

Cl. 13. **Crustacea.** *Ord.* Cirripedia, Siphonostoma, Lophyropoda, Phyllopoda, Pœcilopoda, Læmodipoda, Isopoda, Amphipoda, Stomapoda, Decapoda, Myriapoda.
Cl. 14. **Arachnida.** Orders without names.
Cl. 15. **Insecta.** *a.* **Ametabola.** *Ord.* Aptera. *b.* **Hemimetabola;** *Ord.* Hemiptera, Orthoptera. *c.* **Holometabola.** *Ord.* Diptera, Lepidoptera, Hymenoptera, Strepsiptera, Neuroptera, and Coleoptera.

VERTEBRATA.

VI. Vertebrata.

Cl. 16. **Pisces.** *Subclasses:* 1st. **Leptocardii.** 2d. **Marsipobranchii.** 3d. **Elasmobranchii;** *Ord.* Holocephali, Plagiostomi. 4th. **Ganoidei;** *Ord.* Chrondrostei, Holostei. 5th.

Teleostei; *Ord.* Acanthopteri, Anacanthini, Pharyngognathi, Physostomi, Plectognathi, Lophobranchii. 6th. **Dipnoi.**

Cl. 17. **Reptilia.** *Subclasses:* 1st. **Dipnoa;** *Ord.* Urodela, Batrachia, Gymnophiona. 2d. **Monopnoa:** *a.* Streptostylica; *Ord.* Ophidia, Sauria. *b.* Monimostylica; *Ord.* Chelonia, Crocodila.

Cl. 18. **Aves.**

Cl. 19. **Mammalia.**

The subdivisions of the classes Pisces and Reptilia are taken from the second edition, published in 1854–1856, in which J. Müller's arrangement of the Fishes is adopted; that of the Reptiles is partly Stannius's own. The classes Aves and Mammalia, and the first volume of the second edition, are not yet out.

The most original feature of the classification of von Siebold is the adoption of the types Protozoa and Vermes, in the sense in which they are limited here. The type of Worms has grown out of the investigations of the helminthologists, who, too exclusively engaged with the parasitic Worms, have overlooked their relations to the other Articulata. On the other hand, the isolation in which most entomologists have remained from the zoologists in general has no doubt had its share in preventing an earlier thorough comparison of the Worms and the larval conditions of Insects, without which the identity of type of the Worms, Crustacea, and Insects can hardly be correctly appreciated. Concerning the classes[44] adopted by von Siebold and Stannius I have nothing to remark that has not been said already.

CLASSIFICATION OF R. LEUCKART

The classification of Leuckart is compiled from the following work: Leuckart, *Ueber die Morphologie und die Verwandtschaftsverhältnisse der wirbellosen Thiere* (Brunswick, 1848).

I. Coelenterata, Lkt.

Cl. 1. **Polypi.** *Ord.* Anthozoa and Cylicozoa (Lucernaria.)

Cl. 2. **Acalephæ.** *Ord.* Discophoræ and Ctenophoræ.

II. Echinodermata, Lkt.

Cl. 3. **Pelmatozoa,** Lkt. *Ord.* Cystidea and Crinoidea.

Cl. 4. **Actinozoa,** Latr. *Ord.* Echinida and Asterida.

Cl. 5. **Scytodermata,** Brmst. *Ord.* Holothuriæ and Sipunculida.

III. Vermes.

Cl. 6. **Anenterati,** Lkt. *Ord.* Cestodes and Acanthocephali. (Helminthes, *Burm.*)

Cl. 7. **Apodes,** Lkt. *Ord.* Nemertini, Turbellarii, Trematodes, and Hirudinei. (Trematodes, *Burm.*)

Cl. 8. **Ciliati,** Lkt. *Ord.* Bryozoa and Rotiferi.

Cl. 9. **Annelides.** *Ord.* Nematodes, Lumbricini, and Branchiati. (Annulati, *Burm.*, excl. Nemertinis et Hirudineis.)

[44] The names of the types, Protozoa and Vermes, are older than their limitation in the classification of Siebold. That of Protozoa, first introduced by Goldfuss, has been used in various ways for nearly half a century, while that of Worms was first adopted by Linnæus as a great division of the animal kingdom but in a totally different sense.

IV. Arthropoda.

Cl. 10. **Crustacea.** *Ord.* Entomostraca (Neusticopoda Car.) and Malacostraca.
Cl. 11. **Insecta.** *Ord.* Myriapoda, Arachnida (Acera, Latr.,) and Hexapoda.

V. Mollusca, Cuv. (Palliata, Nitzsch.)

Cl. 12. **Tunicata.** *Ord.* Ascidiæ (Tethyes Sav.) and Salpæ (Thalides Sav.) — Leuckart is somewhat inclined to consider the Tunicata not simply as a class, but even as another great type or branch, intermediate between Echinoderms and Worms.
Cl. 13. **Acephala.** *Ord.* Lamellibranchiata (Cormopoda Nitzsch, Pelecypoda Car.) and Brachiopoda.
Cl. 14. **Gasteropoda.** *Ord.* Heterobranchia (Pteropoda, Inferobranchia, and Tectibranchia), Dermatobranchia (Gymnobranchia and Phlebenterata), Heteropoda, Ctenobranchia, Pulmonata, and Cyclobranchia.
Cl. 15. **Cephalopoda.**

VI. Vertebrata. (Not considered.)

I need not repeat here what I have already stated, in the first section, respecting the primary divisions adopted by Siebold and Leuckart. As to the classes, I may add that his three classes of Echinoderms exhibit only ordinal characters. Besides Birds and Cephalopods, there is not another class so well defined and so little susceptible of being subdivided into minor divisions presenting any thing like class characters as that of Echinoderms. Their systems of organs are so closely homological, that the attempt here made by Leuckart, of subdividing them into three classes, can readily be shown to rest only upon the admission, as classes, of groups which exhibit only ordinal characters, namely, different degrees of complication of structure. With reference to the classes of Worms, the same is equally true, as shown above. The arrangement of these animals proposed by Burmeister is certainly more correct than those of von Siebold and of Leuckart, inasmuch as he refers already correctly the Rotifera to the class of Crustacea and does not, like Leuckart, associate the Bryozoa with the Worms. I agree, however, with Leuckart respecting the propriety of removing the Nemertini and Hirudinei from among the true Annelides. Again, Burmeister appreciates also more correctly the position of the whole type of Worms in referring them, with de Blainville, to the branch of Articulata.

The common fault of all the anatomical classifications which have been proposed since Cuvier consists, first, in having given up to a greater or less extent the fundamental idea of the plan of structure, so beautifully brought forward by Cuvier, and upon which he has insisted with increased confidence and more and more distinct consciousness, ever since 1812; and, second, in having allowed that of complication of structure frequently to take the precedence over the more general features of plan, which, to be correctly appreciated, require, it is true, a deeper insight into the structure of the whole animal kingdom than is needed merely for the investigation of anatomical characters in single types.

Yet if we take a retrospective glance at these systems and especially consider the most recent ones, it must be apparent to those who are conversant with the views now obtaining in our science that, after a test of half a century, the idea of the existence of branches, characterized by different plans of structure as expressing the true relations among animals, has prevailed over the idea of a gradated scale including all animals in one progressive series. When it is considered that this has taken place amidst the most conflicting views respecting Classification and even in the absence of any ruling principle, it must be acknowledged that this can be only owing to the internal truth of the views first propounded by Cuvier. We recognize in the classifications of Siebold, Leuckart, and others the triumph of the great conception of the French naturalist, even though their systems differ greatly from his, for the question whether there are four or more great plans, limited in this or any other way, is not a question of principle but one involving only accuracy and penetration in the investigation; and I maintain that the first sketch of Cuvier, with all its imperfections of details, presents a picture of the essential relations existing among animals truer to nature than the seemingly more correct classifications of recent writers.

SECTION V

PHYSIOPHILOSOPHICAL SYSTEMS

About the time that Cuvier and the French naturalists were tracing the structure of the animal kingdom and attempting to erect a natural system of Zoology upon this foundation, there arose in Germany a school of philosophy under the lead of Schelling which extended its powerful influence to all the departments of physical science. Oken, Kieser, Bojanus, Spix, Huschke,[45] and Carus are the most eminent naturalists who applied the new philosophy to the study of Zoology. But no one identified his philosophical views so completely with his studies in natural history as Oken.

Now that the current is setting so strongly against everything which recalls the German physiophilosophers and their doings, and it has become fashionable to speak ill of them, it is an imperative duty for the impartial reviewer of the history of science to show how great and how beneficial the influence of Oken has been upon the progress of science in general and of Zoology in particular. It is moreover easier, while borrowing his ideas, to sneer at his style and his nomenclature than to discover the true meaning of what is left unexplained in his mostly paradoxical, sententious, or aphoristical expressions; but the man who has changed the whole method of illustrating comparative Osteology — who has carefully investigated the embryology of the higher animals at a time when few physiologists were paying any attention to the subject, who has classified the three kingdoms of nature upon principles wholly his own, who has perceived thousands of homologies and analogies among organized beings entirely overlooked before, who has published an extensive treatise of natural history containing a condensed account of all that was known at the time of its publication, who has conducted for twenty-five years the most extensive and most complete periodical review of the natural sciences ever published, in which every discovery made during a quarter of a century is faithfully recorded, the man who inspired every student with an ardent love for science and with admiration for his teacher — that man will never be forgotten, nor can the services he has rendered to science be overlooked so long as thinking is connected with investigation.

CLASSIFICATION OF OKEN

The following diagram of Oken's classification is compiled from his *Allgemeine Naturgeschichte für alle Stände* (14 vols., Stuttgart, 1833–1843), I. 5. The changes this system has undergone may be ascertained by comparing his *Lehrbuch der Naturphilosophie* (3 vols., Jena, 1809–1811; 2d ed., Jena, 1831; 3rd ed., Zurich, 1843; tr., A. Tulk, London, 1847); *Lehrbuch der Naturgeschichte* (Leipzig, 1812; 2d ed., Weimar, 1815; 3rd ed., Weimar, 1825); *Handbuch der Naturgeschichte zum Gebrach bei Vorlesungen* (2 vols., Nuremberg, 1816–1820); *Naturgeschichte für Schulen* (Leipzig, 1821); and various papers in *Isis* (Jena and Leipzig, 1817–1847).

[45] [Dietrich Kieser, 1779–1862; Ludovici Bojanus, 1776–1827; Emil Huschke, 1797–1858.]

1st Grade. INTESTINAL ANIMALS; also called *Body*-animals and *Touch*-animals. Only one cavity; no head with a brain, only the lowest sense perfect, intestines and skin organs, but no flesh, that is no bones, muscles, or nervous marrow = *Invertebrata.*

Characterized by the development of the vegetative systems of organs, which are those of digestion, circulation, and respiration. Hence —

Cycle I. *Digestive Animals.* = Radiata. Essential character: no development beyond an intestine.
- CL. 1 **Infusoria.** (Stomach animals). Mouth with cilia only, to vibrate.
- CL. 2. **Polypi** (Intestine animals). Mouth with lips and tentacles, to seize.
- CL. 3. **Acalephæ** (Lacteal animals). Body traversed by tubes similar to the lymphatic vessels.

Cycle II. *Circulative Animals.* = Mollusks. Essential character: intestine and vessels.
- CL. 4. **Acephala** (Biauriculate animals). Membranous heart with two auricles.
- CL. 5. **Gasteropoda** (Uniauriculate animals). Membranous heart with one auricle.
- CL. 6. **Cephalopoda** (Bicardial animals). Two hearts.

Cycle III. *Respirative Animals.* = Articulata. Essential character: intestine, vessels, and spiracles.
- CL. 7. **Worms** (Skin animals). Respire with the skin itself, or part of it, no articulated feet.
- CL. 8. **Crustacea** (Branchial animals). Gills or air tubes arising from the horny skin.
- CL. 9. **Insects** (Tracheal animals). Tracheæ internally, gills externally as wings.

2d Grade. FLESH ANIMALS; also called *Head*-animals. = *Vertebrata.* Two cavities of the body, surrounded by fleshy walls (bones and muscles) inclosing nervous marrow and intestines. Head with brain; higher senses developed. Characterized by the development of the animal systems, namely, the skeleton, the muscles, the nerves, and the senses.

Cycle IV. *Carnal Animals proper.* Senses not perfected.
- CL. 10. **Fishes** (*Bone*-animals). Skeleton predominating, very much broken up; muscles white, brain without gyri, tongue without bone, nose not perforated, ear concealed, eyes without lids.
- CL. 11. **Reptiles** (*Muscle*-animals). Muscles red, brain without convolutions, nose perforated, ear without external orifice, eyes immovable with imperfect lids.
- CL. 12. **Birds** (*Nerve*-animals). Brain with convolutions, ears open, eyes immovable, lids imperfect.

Cycle V. *Sensual Animals.* All anatomical systems, and the senses perfected.
- CL. 13. **Mammalia** (*Sense*-animals). Tongue and nose fleshy, ears open, mostly with a conch, eyes movable, with two distinct lids.

The principles laid down by Oken, of which this classification is the practical result for Zoology, may be summed up in the following manner: The grades or great types of Animals are determined by their anatomical systems, such as the body and head; or the intestines and the flesh and senses. Hence two grades in the animal kingdom. Animals are, as it were, the dismembered body of man made alive. The classes of animals are the special representation in living forms of the anatomical systems of the highest being in creation.

Man is considered in this system not only as the key of the whole animal kingdom, but also as the standard measure of the organization of animals. There exists nothing in the animal kingdom which is not represented in higher combinations in Man. The existence of several distinct plans of structure among animals is virtually denied. They are all built after the pattern of Man; the differences among them consist only in their exhibiting either one system only, or a larger or smaller number of systems of organs of higher or lower physiological importance, developed either singly or in connection with one another, in their body. The principles of classification of both Cuvier and Ehrenberg are here entirely negatived. The principle of Cuvier, who admits four different plans of structure in the animal kingdom, is indeed incompatible with the idea that all animals represent only the organs of Man. The principle of Ehrenberg, who considers all animals as equally perfect, is as completely irreconcilable with the assumption that all animals represent an unequal sum of organs; for, according to Oken, the body of animals is, as it were, the analyzed body of Man, the organs of which live singly or in various combinations as independent animals. Each such combination constitutes a distinct class. The principle upon which the orders are founded has already been explained above.

There is something very taking in the idea that Man is the standard of appreciation of all animal structures. But all the attempts which have thus far been made to apply it to the animal kingdom as it exists must be considered as complete failures. In his different works Oken has successively identified the systems of organs of Man with different groups of animals, and different authors, who have adopted the same principle of classification, have identified them in still different ways. The impracticability of such a scheme must be obvious to any one who has satisfied himself practically of the existence of different plans of structure in the organization of animals. Yet the unsoundness of the general principle of the classifications of the physiophilosophers should not render us blind to all that is valuable in their special writings. The works of Oken in particular teem with original suggestions respecting the natural affinities of animals; and his thorough acquaintance with every investigation of his predecessors and contemporaries shows him to have been one of the most learned zoologists of this century.

CLASSIFICATION OF FITZINGER

This diagram is extracted from Fitzinger's *Systema Reptilium* (Vienna, 1843).

I. Provincia. EVERTEBRATA.

Animalia systematum anatomicorum vegetativorum gradum evolutionis exhibentia.

A. Gradus evolutionis systematum physiologicorum vegetativorum.

I. Circulus. GASTROZOA. Evolutio systematis nutritionis.

a. Evolutio prævalens systematis digestionis.	*b.* Evolutio prævalens systematis circulationis.	*c.* Evolutio prævalens systematis respirationis.
CL. 1. **Infusoria.**	CL. 2. **Zoophyta.**	CL. 3. **Acalephæ.**

II. Circulus. PHYSIOZOA. Evolutio systematis generationis.

CL. 4. **Vermes.** CL. 5. **Radiata.** CL. 6. **Annulata.**

B. Gradus evolutionis systematum physiologicorum animalium.

III. Circulus. DERMATOZOA. Evolutio systematis sensibilitatis.

CL. 7. **Acephala.** CL. 8. **Cephalopoda.** CL. 9. **Mollusca.**

IV. Circulus. ARTHROZOA. Evolutio systematis motus.

CL. 10. **Crustacea.** CL. 11. **Arachnoidea.** CL. 12. **Insecta.**

II. Provincia. VERTEBRATA.

Animalia systematum anatomicorum animalium gradum evolutionis exhibentia.

A. Gradus evolutionis systematum physiologicorum vegetativorum.

a. Evolutio systematis nutritionis, simulque ossium: . CL. 13. **Pisces.**
c. Evolutio systematis sensibilitatis, simulque nervorum: CL. 14. **Reptilia.**

B. Gradus evolutionis systematum physiologicorum animalium.

c. Evolutio systematis sensibilitatis, simulque nervorum: CL. 15. **Aves.**
d. Evolutio systematis motus, simulque sensuum: . . CL. 16. **Mammalia.**

The fundamental idea of the classification of Fitzinger is the same as that upon which Oken has based his system. The higher divisions, called by him provinces, grades, and cycles, as well as the classes and orders, are considered as representing either some combination of different systems of organs, or some particular system of organs, or some special organ. His two highest groups (provinces) are the Evertebrata and Vertebrata. The Evertebrata represent the systems of the vegetative organs and the Vertebrata those of the animal organs, as the Gut animals and the Flesh-animals of Oken. Instead, however, of adopting, like Oken, anatomical names for his divisions, Fitzinger employs those most generally in use. His subdivisions or grades of these two primary groups are based upon a repetition of the same differences, within their respective limits. The Invertebrata, in which the vegetative organs prevail, are contrasted with those in which the animal organs prevail, and the same distinction is again drawn among the Vertebrata. Each of these embraces two circles founded upon the development of one particular system of organs, etc. It cannot be expected that the systems founded upon such principles should present a closer agreement with one another than those which are based upon anatomical differences; yet I would ask, what becomes of the principle itself, if its advocates cannot even agree upon what anatomical systems of organs their classes are founded? According to Oken, the Mollusks (Acephala, Gasteropoda, and Cephalopoda) represent the system of circulation; at least in the last edition of his system he views them in that light, whilst Fitzinger considers them as representing the system of sensibility. Oken identifies the Articulata (Worms, Crustacea, and Insects) with the system of respiration, Fitzinger with that of motion, with the exception of the Worms, including Radiata, which he parallelizes with the system of reproduction, etc. Such discrepancies must shake all confidence in these systems, though they should not prevent us from noticing the happy comparisons and suggestions to which the various attempts to classify the animal kingdom in this way have led their authors. It is almost superfluous to add that, great as the disagreement is between the systems of different physiophilosophers, we find quite as striking discrepancies between the different editions of the system of the same author.

The principle of the subdivision of the classes among Invertebrata is here exemplified from the Radiata (Echinodermata). Each series contains three orders.

1st Series.	2d Series.	3d Series.
Evolutio prævalens systematis digestionis.	Evolutio prævalens systematis circulationis.	Evolutio prævalens systematis respirationis.
Asteroidea.	**Echinodea.**	**Scytodermata** (Holothurioids).
1. Encrinoidea.	1. Aprocta.	1. Synaptoidea.
2. Comatulina	2. Echinina.	2. Holothurioidea.
3. Asterina.	3. Spatangoidea.	3. Pentactoidea.

In Vertebrata, each class has five series and each series three orders; so in Mammalia, for example: —

1st Series.	2d Series.	3d Series.	4th Series.	5th Series.
Evolutio prævalens sensus tactus.	Evolutio prævalens sensus gustus.	Evolutio prævalens sensus olfactus.	Evolutio prævalens sensus auditus.	Evolutio prævalens sensus visus.

Cetacea.	Pachydermata.	Edentata.	Unguiculata.	Primates.
1. Balanodea.	1. Phocina.	1. Monotremata.	1. Glires.	1. Chiropteri.
2. Delphinodea.	2. Obesa.	2. Lipodonta.	2. Bruta.	2. Hemipitheci.
3. Sirenia.	3. Ruminantia.	3. Tardigrada.	3. Feræ.	3. Anthropomorphi.

Instead of considering the orders as founded upon a repetition of the characters of higher groups, as Oken would have it, Fitzinger adopts series, as founded upon that idea, and subdivides them further into orders, as above. These series, however, have still less reference to the systems of organs which they are said to represent than either the classes or the higher divisions of the animal kingdom. In these attempts to arrange minor groups of animals into natural series no one can fail to perceive an effort to adapt the frames of our systems to the impression we receive from a careful examination of the natural relations of organized beings. Everywhere we notice such series; sometimes extending only over groups of species, at other times embracing many genera, entire families, nay, extending frequently to several families. Even the classes of the same branch may exhibit more or less distinctly such a serial gradation. But I have failed thus far to discover the principle to which such relations may be referred, as far as they do not rest upon complication of structure[46] or upon the degree of superiority or inferiority of the features upon which the different kinds of groups are themselves founded. Analogy plays also into the series, but before the categories of analogy have been as carefully scrutinized as those of affinity, it is impossible to say within what limits this takes place.

CLASSIFICATION OF MACLEAY

The great merit of the system of MacLeay,[47] and in my opinion it has no other claim to our consideration, consists in having called prominently the attention of naturalists to the difference between two kinds of relationship almost universally confounded before: *affinity* and *analogy*. Analogy is shown to consist in the repetition of

[46] Compare Chap. II, Sect. III.

[47] I have introduced the classification of William Sharpe MacLeay in this section, not because of any resemblance to those of the German physiophilosophers, but on account of its general character, and because it is based upon an ideal view of the affinities of animals.

similar features in groups otherwise remote, as far as their anatomical characters are concerned, whilst affinity is based upon similarity in the structural relations. On account of the similarity of their locomotion, Bats, for instance, may be considered as analogous to Birds; Whales are analogous to Fishes on account of the similarity of their form and their aquatic mode of life; whilst both Bats and Whales are allied to one another and to other Mammalia on account of the identity of the most characteristic features of their structure. This important distinction cannot fail to lead to interesting results. Thus far, however, it has only produced fanciful comparisons from those who first traced it out. It is assumed, for instance, by MacLeay that all animals of one group must be analogous to those of every other group, besides forming a circle in themselves; and in order to carry out this idea all animals are arranged in circular groups in such a manner as to bring out these analogies, whilst the most obvious affinities are set aside to favor a preconceived view. But that I may not appear to underrate the merits of this system, I will present it in the very words of its most zealous admirer and self-complacent expounder, the learned William Swainson.[48]

The *Horæ Entomologicæ,*[49] unluckily for students, can only be thoroughly understood by the adept, since the results and observations are explained in different parts; the style is somewhat desultory, and the groups, for the most part, are rather indicated than defined. The whole, in short, is what it professes to be, more a rough sketch of the leading peculiarities of the great divisions of animals, and the manner in which they are probably connected, than an accurate determination of the groups themselves, or a demonstration of their real affinities. More than this, perhaps, could not have been expected, considering the then state of science, and the herculean difficulties which the author had to surmount. The work in question has now become exceedingly scarce, and this will be an additional reason with us for communicating occasional extracts from it to the reader. Mr. MacLeay's theory will be best understood by consulting his diagram; for he has not, as we have already remarked, defined any of the vertebrated groups. Condensing, however, the result of his remarks, we shall state them as resolvable into the following propositions: 1. That the natural series of animals is continuous, forming, as it were, a circle, so that, upon commencing at any one given point, and thence tracing all the modifications of structure, we shall be imperceptibly led, after passing through numerous forms, again to the point from which we started; 2. That no groups are natural which do not exhibit such a circular series; 3. That the primary divisions of every large group are ten, five of which are composed of comparatively large circles, and

[48] *Geography and Classification of Animals,* pp. 201–205.

[49] MacLeay, *Horæ Entomologicæ, or Essays on the Annulose Animals* (2 vols., London, 1819–1821).

five of smaller: these latter being termed osculant, and being intermediate between the former, which they serve to connect; 4. That there is a tendency in such groups as are placed at the opposite points of a circle of affinity "to meet each other;" 5. That *one* of the five larger groups into which every natural circle is divided, "bears a resemblance to all the rest, or, more strictly speaking, consists of types which represent those of each of the four other groups, together with a type peculiar to itself." These are the chief and leading principles which Mr. MacLeay considers as belonging to the natural system. We shall now copy his diagram, or table of the animal kingdom, and then endeavor, with this help, to explain the system more in detail.

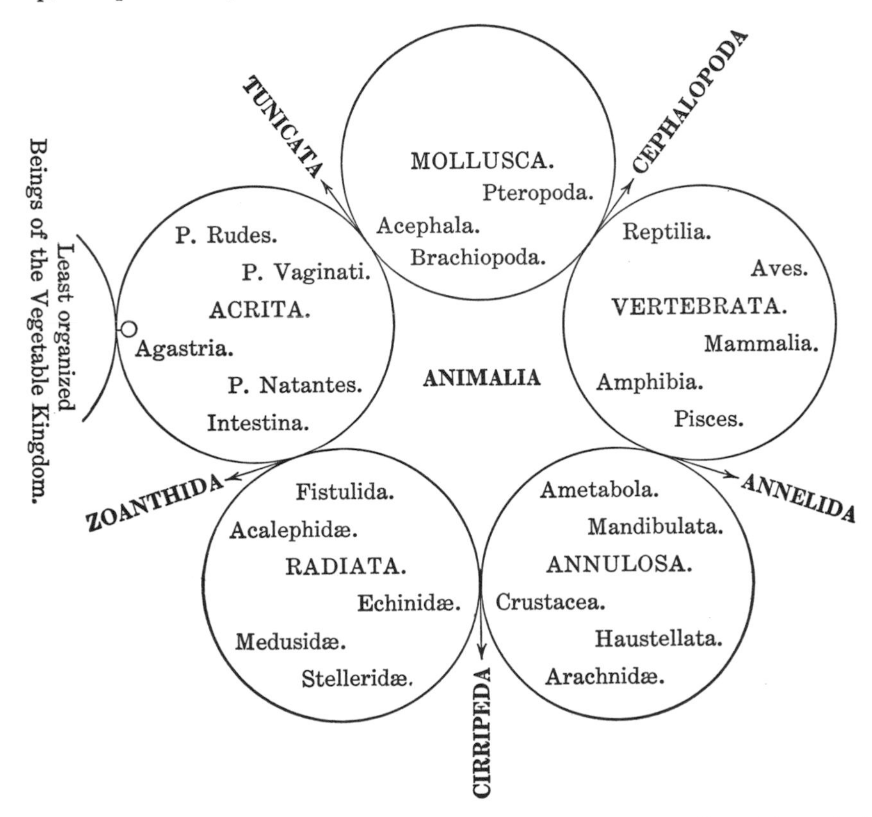

We must, in the first instance, look to the above tabular disposition of all animals, as forming themselves collectively into one great circle, which circle touches or blends into another, composed of plants, by means of the "least organized beings of the vegetable kingdom." Next we are to look to the larger component parts of this great circular assemblage. We find it, in accordance with the third proposition, to exhibit five great circles, composed of the MOLLUSCA, or shellfish; ACRITA, or polyps; RADIATA, or star-fish; ANNULOSA, or insects; and VERTEBRATA, or vertebrated animals; each passing or blending into each other, by means of five other groups of animals, much smaller, indeed, in their extent, but forming so many connecting or osculant circles.[50] The number, therefore, as many erroneously suppose, is not five, but ten. This is quite obvious; and our opinion on this point is confirmed by the author himself, in the following passage, when alluding to his remarks upon the whole: "The foregoing observations, I am well aware, are far from accurate, but they are sufficient to prove that there are five great circular groups in the animal kingdom, each of which possesses a peculiar structure; and that these, when connected by means of five smaller osculant groups, compose the whole province of Zoology." Now these smaller osculant groups are to be viewed as circles, for, as it is elsewhere stated, "every natural group is a circle, more or less complete." This, in fact, is the third general principle of Mr. MacLeay's system, and he has exemplified his meaning of a natural group in the above diagram, where all animals are arranged under five large groups or circles, and five smaller ones. Let us take one of these groups, the Vertebrata: does that form a circle of itself? Yes; because it is intimated that the Reptiles (*Reptilia*) pass into the Birds (*Aves*), these again into the Quadrupeds (*Mammalia*). Quadrupeds unite with the Fishes (*Pisces*) these latter with the amphibious Reptiles, and the Frogs bring us back again to the Reptiles, the point from whence we started. Thus, the series of the vertebrated group is marked out and shown to be circular; therefore, it is a natural group. This is an instance where the circular series can be traced. We now turn to one where the series is imperfect, but where there is a decided tendency to a circle: this is the Mollusca. Upon this group our author says, "I have by no means determined the circular disposition to hold good among the Mollusca; still, as it is equally certain that this group of animals is as yet the least known, it may be improper, at present, to conclude that it forms any exception to the rule; it would even seem unquestionable that the Gasteropoda of Cuvier return into themselves, so as to form a circular group; but whether the Acephala form one or two such, is by no means accurately ascertained, though enough is known of the Mollusca to incline us to suspect that they are no less subjected, in general, to a circular disposition than the four other great groups." This, therefore, our author considers as one of those groups which, without actually forming a circle, yet evinces a disposition to do so; and it is therefore presumed to be a natural group. But, to illustrate this principle farther, let us return to the circle of Vertebrata. This, as we see by the diagram, contains five minor groups, or circles, each of which is again resolvable into five others, regulated precisely in the same way. The class Aves, for example, is first divided into rapacious birds (*Raptores*), perching birds (*Insessores*), gallinaceous birds (*Rasores*), wading birds (*Grallatores*), and swimming birds (*Natatores*); and the proof of this class being a natural group is in all these divisions blending into each other at their confines and forming a circle. In this manner we proceed, beginning with the higher groups, and descending to the lower, until at length we descend to genera, properly so called, and reach, at last, the species; every group, whether large or small, forming a circle of its own. Thus there are circles within circles, "wheels within wheels," — an infinite number of complicated relations; but all regulated by one simple and uniform principle, — that is, the circularity of every group.

[50] In the original diagram, as in that above, these five smaller circles are not represented graphically, but merely indicated by the names arranged like rays between the five large circles.

The writer who can see that the Quadrupeds unite with the Fishes and the like and yet says that Cuvier "was totally unacquainted with the very first principles of the natural system," hardly deserves to be studied in our days.

The attempt at representing graphically the complicated relations which exist among animals has, however, had one good result; it has checked, more and more, the confidence in the uniserial arrangement of animals and led to the construction of many valuable maps exhibiting the multifarious relations which natural groups of any rank bear to one another.

SECTION VI

EMBRYOLOGICAL SYSTEMS

Embryology, in the form it has assumed within the last fifty years, is as completely a German science as the "Naturphilosophie." It awoke to this new activity contemporaneously with the development of the Philosophy of Nature. It would hardly be possible to recognize the leading spirit in this new development, from his published works; but the man whom Pander and K. E. von Baer acknowledge as their master must be considered as the soul of this movement, and this man is Ignatius Döllinger.[51] It is with deep gratitude I remember for my own part the influence that learned and benevolent man had upon my studies and early scientific application during the four years I spent in his house in Munich, from 1827 to 1831; to him I am indebted for an acquaintance with what was then known of the development of animals prior to the publication of the great work of von Baer; and from his lectures I first learned to appreciate the importance of Embryology to Physiology and Zoology. The investigations of Pander[52] upon the development of the chicken in the egg, which have opened the series of those truly original researches in Embryology of which Germany may justly be proud, were made under the direction and with the cooperation of Döllinger, and were soon followed by the more extensive works of Rathke and von Baer,

[51] [1770–1841.]

[52] Heinrich C. von Pander, *Beiträge zur Entwickelungsgeschichte des Hühnchens im Eie* (Würzburg, 1817).

whom the civilized world acknowledges as the founders of modern Embryology.

The principles of classification propounded by K. E. von Baer seem never to have been noticed by systematic writers, and yet they not only deserve the most careful consideration, but it may fairly be said that no naturalist besides Cuvier has exhibited so deep an insight into the true character of a natural system, supported by such an extensive acquaintance with the subject, as this great embryologist has in his "Scholien und Corallarien zu der Entwickelungsgeschichte des Hühnchens im Eie."[53] These principles are presented in the form of general proportions, rather than in the shape of a diagram with definite systematic names, and this may explain the neglect which it has experienced on the part of those who are better satisfied with words than with thoughts. A few abstracts, however, may show how richly the perusal of his work is likely to reward the reader.

The results at which K. E. von Baer had arrived by his embryological investigations respecting the fundamental relations existing among animals differed considerably from the ideas then prevailing. In order therefore to be correctly understood, he begins with his accustomed accuracy and clearness to present a condensed account of those opinions with which he disagreed, in these words: —

> Few views of the relations existing in the organic world have received so much approbation as this: that the higher animal forms, in the several stages of the development of the individual, from the beginning of its existence to its complete formation, correspond to the permanent forms in the animal series, and that the development of the several animals follows the same laws as those of the entire animal series; that consequently the more highly organized animal, in its individual development, passes in all that is essential through the stages that are permanent below it, so that the periodical differences of the individual may be reduced to the differences of the permanent animal forms.

Next, in order to have some standard of comparison with his embryological results, he discusses the relative position of the different permanent types of animals, as follows: —

> It is specially important that we should distinguish between the degree of perfection in the animal structure and the type of organization. The degree of perfection of the animal structure consists in the greater or less heterogeneous-

[53] See von Baer, *Ueber Entwickelungsgeschichte der Thiere* (1828). See also his "Beiträge zur Kenntniss der niedern Thiere," *Nova Acta Academiæ Caesareæ Leopoldino-Carolinæ,* XIII (1827), 525–762, and "Ueber das aüssere und innere Skeleton," *Archiv für Anatomie und Physiologie v. J. F. Meckel,* I (1826), 327–376.

ness of the elementary parts, and the separate divisions of a complicatd apparatus, — in one word, in the greater histological and morphological differentiation. The more uniform the whole mass of the body is, the lower the degree of perfection; it is a stage higher when nerve and muscle, blood and cellular tissue, are sharply distinguished. In proportion to the difference between these parts, is the development of the animal life in its different tendencies; or, to express it more accurately, the more the animal life is developed in its several tendencies, the more heterogeneous are the elementary parts which this life brings into action. The same is true of the single parts of any apparatus. That organization is higher in which the separate parts of an entire system differ more among themselves, and each part has greater individuality, than that in which the whole is more uniform. I call type, the relations of organic elements and organs, as far as their position is concerned. This relation of position is the expression of certain fundamental connections in the tendency of the individual relations of life; as, for instance, of the receiving and discharging poles of the body. The type is altogether distinct from the degree of perfection, so that the same type may include many degrees of perfection, and *vice versâ,* the same degree of perfection may be reached in several types. The degree of perfection, combined with the type, first determines those great animal groups which have been called classes.[54] The confounding of the degree of perfection with the type of organization seems the cause of much mistaken classification, and in the evident distinction between these two relations we have sufficient proof that the different animal forms do not present one uniserial development, from the Monad up to Man.

The types he has recognized are: —

I. *The Peripheric Type.* The essential contrasts in this type are between the centre and the periphery.[55] The organic functions of life are carried on in antagonistic relations from the centre to the circumference. Corresponding to this, the whole organization radiates around a common centre. There exists besides only the contrast between above and below, but in a weaker degree; that between right and left, or before and behind, is not at all noticeable, and the motion is therefore undetermined in its direction. As the whole organization radiates from one focus, so are the centres of all the organic systems arranged, ring-like, around it, as, for instance, the stomach, the nerves and vessels (if these parts are developed) and the branches

[54] From this statement it is plain that von Baer has a very definite idea of the plan of structure and that he has reached it by a very different road from that of Cuvier. It is clear also that he understands the distinction between a plan and its execution. But his ideas respecting the different features of structure are not quite so precise. He does not distinguish, for instance, between the complication of structure as determining the relative rank of the orders, and the different ways in which, and the different means with which the plans are executed, as characteristic of the classes.

[55] Without translating verbatim the descriptions von Baer gives of his types, which are greatly abridged here, they are reproduced as nearly as possible in his own words.

extending from them into the rays. What we find in one ray is repeated in every other, the radiation being always from the centre outwards, and every ray bearing the same relation to it.

II. *The Longitudinal Type,* as observed in the Vibrio, the Filaria, the Gordius, the Nais, and throughout the whole series of articulated animals. The contrast between the receiving and the discharging organs, which are placed at the two ends of the body, controls the whole organization. The mouth and the anus are always at opposite ends, and usually also the sexual organs, though their opening is sometimes farther forward; this occurs, however, more frequently in the females, in which these organs have a double function, than in the males. When both sexual organs are removed from the posterior extremity, the opening in the female usually lies farther forward than in the male. So is it in the Myriapods and the Crabs. The Leeches and Earthworms present a rare exception. The receptive pole being thus definitely fixed, the organs of senses, as instrumental to the receptivity of the nervous system, early reach an important degree of perfection. The intestinal canal, as well as the vascular stems of the nervous system, extend through the whole length of the body, and all organic motion in these animals has the same prevailing direction. Only subordinate branches of these organs arise laterally, and chiefly wherever the general contrast, manifested in the whole length, is repeated in such a manner that, for each separate segment, the same contrast arises anew in connection with the essential elements of the whole organism. Hence the tendency in these animals to divide into many segments in the direction of the longitudinal axis of the body. In the true Insects undergoing metamorphosis, these segments unite again into three principal regions, in the first of which the life of the nerves prevails; in the second, motion; in the third, digestion; though neither of the three regions is wholly deprived of any one of these functions. Besides the opposition between before and behind, a less marked contrast is observed in a higher stage of development between above and below. A difference between right and left forms a rare exception and is generally wanting. Sensibility and irritability are particularly developed in this series. Motion is active and directed more decidedly forward in proportion as the longitudinal axis prevails. When the body is contracted as in spiders and crabs, its direction is less decided. The plastic organs are little developed; glands, especially, are rare, and mostly replaced by simple tubes.

III. *The Massive Type.* We may thus call the type of Mollusks, for neither length nor surface prevails in them, but the whole body and its separate parts are formed rather in round masses which may be either hollow or solid. As the chief contrast of their structure is not between the opposite ends of the body, nor between the centre and periphery, there is almost throughout this type an absence of symmetry. Generally the discharging pole is to the right of the receptive one. The discharging pole, however, is either near the receptive one or removed from it and approximated to the posterior extremity of the body. As the tract of the digestive apparatus is always determined by these two poles, it is more or less arched; in its simplest form it is only a single arch, as in Plumatella. When that canal is long, it is curled up in a spiral in the centre, and the spiral probably has its definite laws. For instance, the anterior part of the alimentary canal appears to be always placed under the posterior. The principal currents of blood are also in arches which do not coincide with the medial line of the body. The nervous system consists of diffused ganglia united by threads, the larger ones being around the œsophagus. The nervous system and the organs of sense appear late; the motions are slow and powerless.

IV. *The Vertebrate Type.* This is, as it were, composed of the preceding types, as we distinguish an animal and a vegetative system of the body, which, though influencing one another in their development, have singly a peculiar typical organization. In the animal system the articulation reminds us of the second type, and the discharging and receiving organs are also placed at opposite ends. There is, however, a marked difference between the Articulates and the Vertebrates, for the animal system of the Vertebrates is not only doubled along the two sides, but at the same time upwards and downwards, in such a way that the two lateral walls which unite below circumscribe the vegetative system, while the two tending upward surround a central organ of the animal life, the brain and spinal marrow, which is wanting in Invertebrates. The solid frame represents this type most completely, as from its medial axis, the backbone, there arise upward arches which close in an upper crest and downward arches which unite, more or less, in a lower crest. Corresponding to this we see four rows of nervous threads along the spinal marrow, which itself contains four strings, and a quadripartite grey mass. The muscles of the trunk form also four principal masses, which are particularly distinct in the Fishes. The animal system is therefore doubly symmetrical in its arrangement. It might easily be shown how the vegetative systems of the body correspond to the type of Mollusks, though influenced by the animal system.

From the illustrations accompanying this discussion of the great types or branches of the animal kingdom, and still more from the paper published by K. E. von Baer in the *Nova Acta,*[56] it is evident that he perceived more clearly and earlier than any other naturalist the true relations of the lowest animals to their respective branches. He includes neither Bryozoa nor Intestinal Worms among Radiata, as Cuvier and after him so many modern writers did, but correctly refers the former to the Mollusks and the latter to the Articulates.

Comparing these four types with the embryonic development, von Baer shows that there is only a general similarity between the lower animals and the embryonic stages of the higher ones, arising mainly from the absence of differentiation in the body, and not from a typical resemblance. The embryo does not pass from one type to the other; on the contrary, the type of each animal is defined from the beginning and controls the whole development. The embryo of the Vertebrate is a Vertebrate from the beginning and does not exhibit at any time a correspondence with the Invertebrates. The embryos of Vertebrates do not pass in their development through other permanent types of animals. The fundamental type is first developed, afterwards more and more subordinate characters appear. From a more general type the more special is manifested, and the more two forms of animals differ, the earlier must their development be traced back to discern an agreement between them. It is barely possible that in their first beginning all animals are alike and present only hollow spheres, but the individual development of the higher animals certainly does not pass through the permanent forms of lower

[56] "Beiträge zur Kenntniss . . . ," *Nova Acta* . . . , XIII (1827), containing seven papers, upon Aspidogaster, Distoma, Cercaria, Nitzchia, Polystoma, Planaria, and the general affinities of all animals. These "Beiträge" and the papers in which Cuvier characterized for the first time the four great types of the animal kingdom (*Annales du muséun* . . . , XIX, 1812), are among the most important contributions to general Zoology ever published.

ones. What is common in a higher group of animals is always sooner developed in their embryos than what is special; out of that which is most general arises that which is less general, until that which is most special appears. Each embryo of a given type of animals, instead of passing through other definite types, becomes on the contrary more and more unlike them. An embryo of a higher type is, therefore, never identical with another animal type, but only with an embryo.[57]

Thus far do the statements of von Baer extend.[58] It is evident from this that he has clearly perceived the limitation of the different modes of embryonic development within the respective branches of the animal kingdom, but it is equally certain that his assertions are too general to furnish a key for the comparison of the successive changes which the different types undergo within their respective limits, and that he is still vaguely under the impression that the development corresponds in its individualization to the degrees of complication of structure. This could hardly be otherwise, as long as the different

[57] [This is a succinct and entirely accurate statement of von Baer's views on the recapitulation concept, which Darwin would have profited from had he been fully aware of the problem. For a full exposition of the relationship of Darwin, von Baer, and Agassiz to the recapitulation idea, see Jane Oppenheimer, "An Embryological Enigma in the *Origin of Species*," in Bentley Glass, *et al.* (eds.), *Forerunners of Darwin, 1745–1859* (Baltimore, 1959), pp. 292–322.]

[58] The account which Huxley gives of von Baer's views, (see Powell, *Essays*, Appendix 7, p. 495) is incorrect. Von Baer did not "demonstrate that the classification of Cuvier was, in the main, simply the expression of the fact, that there are certain common *plans of development* in the animal kingdom," etc., for Cuvier recognized these plans in the *structure* of the animals, before von Baer traced their development, and von Baer himself protests against an identification of his views with those of Cuvier (*Entwickelungsgeschichte*, I, 7). Nor has von Baer demonstrated the "doctrine of the unity of organization of all animals" and placed it "upon a footing as secure as the law of gravitation," and arrived at "the grandest law" that, up to a certain point, the development *"followed a plan common to all animals."* On the contrary, von Baer admits four distinct types of animals and four modes of development. He only adds: "It is barely possible that in their first beginning all animals are alike." Huxley must also have overlooked Cuvier's introduction to the *Règne animal,* (2d ed.), I, p. 48, when he stated that Cuvier "did not attempt to discover upon what plans animals are constructed, but to ascertain in what manner the facts of animal organizations could be thrown into the fewest possible propositions." On the contrary, Cuvier's special object for many years has been to point out these plans and to show that they are characterized by peculiar structures, while von Baer's merit consists in having discovered four *modes of development* which coincide with the branches of the animal kingdom, in which Cuvier recognized four different *plans of structure*. Huxley is equally mistaken when he says that Cuvier adopted the nervous system "as the base of his great divisions."

categories of the structure of animals had not been clearly distinguished.[59]

CLASSIFICATION OF K. E. VON BAER

In conformity with his embryological investigations, K. E. von Baer proposes the following classification.

I. Peripheric Type. (RADIATA.) Evolutio radiata. The development proceeds from a centre, producing identical parts in a radiating order.

II. Massive Type. (MOLLUSCA.) Evolutio contorta. The development produces identical parts curved around a conical or other space.

III. Longitudinal Type. (ARTICULATA.) Evolutio gemina. The development produces identical parts arising on both sides of an axis and closing up along a line opposite the axis.

IV. Doubly Symmetrical Type. (VERTEBRATA.) Evolutio bigemina. The development produces identical parts arising on both sides of an axis, growing upwards and downwards, and shutting up along two lines, so that the inner layer of the germ is inclosed below and the upper layer above. The embryos of these animals have a dorsal cord, dorsal plates, and ventral plates, a nervous tube and branchial fissures.

- 1°. They acquire branchial fringes;
  - *a*. But no genuine lungs are developed.
    - α. The skeleton is not ossified. **Cartilaginous Fishes.**
    - β. The skeleton is ossified. **Fishes proper.**
  - *b*. Lungs are formed. **Amphibia.**
    - α. The branchial fringes remain. **Sirens.**
    - β. The branchial fringes disappear. **Urodela** and **Anura.**
- 2°. They acquire an allantois, but
  - *a*. Have no umbilical cord;
    - α. Nor wings and air sacs. **Reptiles.**
    - β. But wings and air sacs. **Birds.**
  - *b*. Have an umbilical cord. **Mammalia.**
    - α. Which disappears early;
      - 1°. Without connection with the mother. **Monotremata.**
      - 2°. After a short connection with the mother. **Marsupialia.**
    - β. Which is longer persistent;
      - 1°. The yolk sac continues to grow for a long time.
        - The allantois grows little. **Rodentia.**
        - The allantois grows moderately. **Insectivora.**
        - The allantois grows much. **Carnivora.**
      - 2°. The yolk sac increases slightly.
        - The allantois grows little. Umbilical cord very long. **Monkeys** and **Man.**
        - The allantois continues to grow for a long time. Placenta in simple masses. **Ruminants.**
        - The allantois continues to grow for a long time. Placenta spreading. **Pachyderms** and **Cetacea.**

CLASSIFICATION OF VAN BENEDEN

Pierre J. Van Beneden has also proposed a classification based upon Embryology, which was first sketched in his paper upon the Embryology of Bryozoa: *Recherches sur l'anatomie, la physiologie et l'embryogénie des Bryozoaires . . .* (Brussels, 1845) (pamphlet), and afterwards extended in his *Anatomie comparée* (Brussels, without date, but probably from the year 1855).

[59] Compare Chap. II, Sect. I to IX.

I. Hypocotyledones or Hypovitellians. (Vertebrata.) The vitellus enters the body from the ventral side.

Cl. 1. **Mammalia.** (Primates, Cheiroptera, Insectivora, Rodentia, Carnivora, Edentata, Proboscidea, Ungulata, Sirenoidea, Cetacea.)

Cl. 2. **Birds.** (Psittaceæ, Rapaces, Passeres, Columbæ, Gallinæ, Struthiones, Grallæ, Palmipedes.)

Cl. 3. **Reptiles.** (Crocodili, Chelonii, Ophidii, Saurii, Pterodactyli, Simosauri, Plesiosauri, Ichthyosauri.)

Cl. 4. **Batrachians.** (Labyrinthodontes, Peromelia, Anura, Urodela, Lepidosirenia.)

Cl. 5. **Fishes.** (Plagiostomi, Ganoidei, Teleostei, Cyclostomi, Leptocardii.)

II. Epicotyledones or Epivitellians. (Articulata.) The vitellus enters the body from the dorsal side.

Cl. 6. **Insects.** (Coleoptera, Nevroptera, Strepsiptera, Hymenoptera, Lepidoptera, Diptera, Orthoptera, Hemiptera, Thysanura, Parasita.)

Cl. 7. **Myriapodes.** (Diplopoda, Chilopoda.)

Cl. 8. **Arachnides.** (Scorpiones, Araneæ, Acari, Tardigrada.)

Cl. 9. **Crustacea.** (Decapoda, Stomapoda, Amphipoda, Isopoda, Læmodipoda, Phyllopoda, Lophyropoda, Xiphosura, Siphonostma, Myzostoma, and Cirripedia.)

III. Allocotyledones or Allovitellians. (Mollusco-Radiaria.) The vitellus enters the body neither from the ventral nor from the dorsal side.

Cl. 10. **Mollusca.** Including Cephalopoda, Gasteropoda, Pœcilopoda, and Brachiopoda. (Acephala, Tunicata, and Bryozoa.)

Cl. 11. **Worms.** (Malacopoda, Annelides, Siponculides, Nemertini, Nematodes, Acanthocephali, Scoleides, Hirudinei.)

Cl. 12. **Echinoderms.** (Holothuriæ, Echinides, Stellerides, Crinoides, Trematodes, Cestodes, Rotiferi, Planariæ.)

Cl. 13. **Polyps.** Including Tunicata, Bryozoa, Anthozoa, Alcyonaria, and Medusæ, as orders. (Ctenophoræ, Siphonophoræ, Discophoræ, Hydroids, Anthophoridæ.)

Cl. 14. **Rhizopods.** Only the genera mentioned.

Cl. 15. **Infusoria.** Only genera and families mentioned.

Van Beneden thinks the classification of Linnæus truer to nature than either that of Cuvier or of de Blainville, as the class of Worms of the Swedish naturalist corresponds to his Allocotyledones, that of Insects to his Hypocotyledones, and the four classes of Pisces, Amphibia, Aves, and Mammalia to his Hypocotyledones. He compares his primary divisions to the Dicotyledones, Monocotyledones, and Acotyledones of the vegetable kingdom. But he overlooks that the Cephalopods are not Allocotyledones, and that any group of animals which unites Mollusks, Worms, and Radiates in one great mass cannot be founded upon correct principles. As to his classes, I can only say that if there are natural classes among animals, there never was a combination of animals proposed since Linnæus less likely to answer to a philosophical idea of what a class may be than that which unites Tunicata with Polyps and Acalephs. In his latest work Van Beneden has introduced in this classification many important improvements and additions. Among the additions the indication of the orders, which are introduced in brackets in the diagram above, deserve to be particularly noticed. These changes relate chiefly to the Mollusks and Polyps; the Tunicata and Bryozoa being removed from the Polyps to the Mollusks. The Acalephs and Polypi, however, are still considered as forming together one single class.

The comparison instituted by Van Beneden between his classification of the animal kingdom and that of the plants most generally adopted now, leads me to call again attention to the necessity of carefully scrutinizing anew the vegetable kingdom, with the view of ascertaining how far the results I have arrived at concerning the value of the different kinds of natural groups existing among animals apply also to the plants. It would certainly be premature to assume that because the branches of the animal kingdom are founded upon different plans of structure the vegetable kingdom must necessarily be built also upon different plans. There are probably not so many different modes of development among plants as among animals; unless the reproduction by spores, by naked polyembryonic seeds, by angiospermous monocotyledonous seeds, and by angiospermous dicotylodonous seeds, connected with the structural differences exhibited by the Acotyledones, Gymnospermes, Monocotyledones, and Dicotyledones, be considered as amounting to an indication of different plans of structure. But even then these differences would not be so marked as those which distinguish the four branches of the animal kingdom. The limitation of classes and orders, which presents comparatively little difficulty in the animal kingdom, is least advanced among plants, whilst botanists have thus far been much more accurate than zoologists in characterizing families. This is no doubt chiefly owing to the peculiarities of the two organic kingdoms.

It must be further remarked that in the classification of Van Beneden the animals united under the name of Allocotyledones are built upon such entirely different plans of structure, that their combination should of itself satisfy any unprejudiced observer that any principle which unites them in that way cannot be true to nature.

DIAGRAM OF THE DEVELOPMENT OF ANIMALS BY KÖLLIKER

KÖLLIKER in his *Entwickelungsgeschichte der Cephalopoden* (Zurich, 1844), I, 175, has submitted the following diagram of the development of the animal kingdom.

A. The embryo arises from a primitive part. (Evolutio ex una parte.)
  1°. It grows in two directions, with bilateral symmetry. (Evolutio bigemina.)
    a. The dorsal plates close up. **Vertebrata.**
    b. The dorsal plates remain open and are transformed into limbs. **Articulata.**
  2°. It grows uniformly in every direction. (Evolutio radiata.) And
    a. Incloses the embryonal vesicle entirely.
      α. This takes place very early. **Gasteropoda** and **Acephala.**
      β. This takes place late. (Temporary vitelline sac.) **Limax.**
    b. Contracts above the embryonal vesicle. (Genuine vitelline sac.) **Cephalopoda.**

B. The whole body of the embryo arises simultaneously. (Evolutio ex omnibus partibus.)
  1°. It grows in the direction of its transverse axis,
    a. With its hind body. **Radiata.** (Echinoderms.)
    b. With the fore body, and
      α. The hind body does not grow. **Acalephs.**
      β. The hind body grows longitudinally. **Polypi.**
  2°. It grows in the direction of its longitudinal axis. **Worms.**

I have already shown how unnatural a zoological system must be which is based upon a distinction between total or partial segmentation of the yolk. No more can a diagram of the development of animals which adopts this difference as fundamental be true to nature, even though it is based upon real facts. We ought never to single out isolated features by which animals may be united or separated, as most anatomists do; our aim should rather be to ascertain their general relations, as Cuvier and K. E. von Baer have so beautifully shown. I think also that the homology of the limbs of Articulata and the dorsal plates of Vertebrata is more than questionable. The distinction introduced between Polyps and Acalephs and these and the other Radiates is not any better founded. It seems also quite inappropriate to call the development of Mollusks evolutio radiata, especially after Baer had designated under that same name the mode of formation of the branch of Radiates, for which it is far better adapted.

The classification of Vogt[60] presents several new features, one of which is particularly objectionable. I mean the separation of the Cephalopoda from the other Mollusks, as a distinct primary division of the animal kingdom. Having adopted the fundamental distinction introduced by Kölliker between the animals in which the embryo is developed from the whole yolk, and those in which it arises from a distinct part of it, Vogt was no doubt led to this step in consequence of his interesting investigations upon Actæon, in which he

[60] *Zoologische Briefe;* see above, n. 3.

CLASSIFICATION OF VOGT

Contrast between the Embryo and the Yolk.

I. VERTEBRATA. Yolk ventral.

CL. 1. **Mammalia.** 1°. **Aplacentaria;** *Ord.* Monotremata, Marsupialia. 2°. **Placentaria.** Ser. 1. *Ord.* Cetacea, Pachydermata, Solidungula, Ruminantia, and Edentata; *S.* 2. Pinnipedia, Carnivora; *S.* 3. Insectivora, Volitantia, Glires, Quadrumana, Bimana.
CL. 2. **Aves.** *Ser.* 1. Insessores; *Ord.* Columbæ, Oscines, Clamatores, Scansores, Raptatores; *Ser.* 2. Autophagi; *Ord.* Natatores, Grallatores, Gallinacea, Cursores.
CL. 3. **Reptilia.** *Ord.* Ophidia, Sauria, Pterodactylia, Hydrosauria, and Chelonia.
CL. 4. **Amphibia.** *Ord.* Lepidota, Apoda, Caudata, Anura.
CL. 5. **Pisces.** *Ord.* Leptocardia, Cyclostomata, Selachia, Ganoidea, Teleostia.

II. ARTICULATA. Yolk dorsal.

CL. 6. **Insecta.** *Subcl.* 1. **Ametabola;** *Ord.* Aptera. *Subcl.* 2. **Hemimetabola;** *Ord.* Hemiptera and Orthoptera. *Subcl.* 3. **Holometabola;** *Ord.* Diptera, Lepidoptera, Strepsiptera, Neuroptera, Coleoptera, Hymenoptera.
CL. 7. **Myriapoda.** Only divided into families.
CL. 8. **Arachnida.** *Series* 1. Pycnogonida and Tardigrada; *Ord.* Acarina, Araneida. *Series* 2. With three families.
CL. 9. **Crustacea.** *Subcl.* 1. **Entomostraca;** *Ord.* Cirripedia, Parasita, Copepoda, Phyllopoda, Trilobita, Ostracoda. *Subcl.* 2. **Xiphosura.** *Subcl.* 3. **Podophthalma;** *Ord.* Stomapoda, Decapoda. *Subcl.* 4. **Edriophthalma;** *Ord.* Læmipoda, Amphipoda, Isopoda.

Transformation of the whole Yolk into the Embryo.

III. CEPHALOPODA. Yolk cephalic.

CL. 10. **Cephalopoda.** *Ord.* Tetrabranchiata and Dibranchiata.

IV. MOLLUSCA. Irregular disposition of organs.

CL. 11. **Cephalophora.** *Subcl.* 1. **Pteropoda.** *Subcl.* 2. **Heteropoda.** *Subcl.* 3. **Gasteropoda;** *Ord.* Branchiata and Pulmonata. — Chitonida.
CL. 12. **Acephala.** *Subcl.* 1. **Brachiopoda;** *Ord.* Rudista, Brachiopoda. *Subcl.* 2. **Lamellibranchia;** *Ord.* Pleurochoncha, Orthoconcha, Inclusa.
CL. 13. **Tunicata.** *Ord.* Ascidiæ, Biphora. } Molluscoidea.
CL. 14. **Ctenophora.** Only subdivided into families. } Molluscoidea.
CL. 15. **Bryozoa.** *Ord.* Stelmatopoda, Lophopoda. } Molluscoidea.

V. VERMES. Organs bilateral.

CL. 16. **Annelida.** *Ord.* Hirudinea, Gephyrea, Scoleina, Tubicola, Errantia.
CL. 17. **Rotatoria.** *Ord.* Sessilia, Natantia.
CL. 18. **Platyelmia.** 1°. *Ord.* Cestoidea, Trematoda. 2°. *Ord.* Planarida, Nemertina.
CL. 19. **Nematelmia.** *Ord.* Gregarinea, Acanthocephala, Gordiacei, Nematoidei.

VI. RADIATA. Organs radiate.

CL. 20. **Echinodermata.** *Ord.* Crinoidea, Stellerida, Echinida, Holothurida.
CL. 21. **Siphonophora.** Only subdivided into families.
CL. 22. **Hydromedusæ.** Not clearly subdivided into orders.
CL. 23. **Polypi.** *Ord.* Hexactinia, Pentactinia, Octactinia.

No Egg.

VII. PROTOZOA.

CL. 24. **Infusoria.** *Ord.* Astoma and Stomatoda.
CL. 25. **Rhizopoda.** *Ord.* Monosomatia and Polythalamia.

found a relation of the embryo to the yolk differing greatly from that observed by Kölliker in Cephalopods. But as I have already shown above, this cannot any more justify their separation as branches than the total segmentation of the yolk of Mammalia could justify the separation of the latter from the other Vertebrates. Had the distinction made by Vogt between Cephalopods and the other Mollusks the value he assigns to it, Limax should also be separated from the other Gasteropods. The assertion that Protozoa produce no eggs deserves no special consideration after what has already been said in the preceding sections respecting the animals themselves. As to the transfer of the Ctenophora to the type of Mollusks, it can in no way be maintained.

Before closing this sketch of the systems of Zoology, I cannot forego the opportunity of adding one general remark. If we remember how completely independent the investigations of K. E. von Baer were from those of Cuvier, how different the point of view was from which they treated their subject, the one considering chiefly the mode of development of animals, while the other looked mainly to their structure; if we further consider how closely the general results at which they have arrived agree throughout, it is impossible not to be deeply impressed with confidence in the opinion they both advocate, that the animal kingdom exhibits four primary divisions, the representatives of which are organized upon four different plans of structure and grow up according to four different modes of development. This confidence is further increased when we perceive that the new primary groups which have been proposed since are neither characterized by such different plans, nor developed according to such different modes of development, but exhibit simply minor differences. It is indeed a very unfortunate tendency, which prevails now almost universally among naturalists with reference to all kinds of groups, of whatever value they may be, from the branches down to the species, to separate at once from one another any types which exhibit marked differences, without even inquiring first whether these differences are of a kind that justifies such separations. In our systems the quantitative element of differentiation prevails too exclusively over the qualitative. If such distinctions are introduced under well-sounding names they are almost certain to be adopted; as if science gained anything by concealing a difficulty under a Latin or Greek name, or was advanced by the additional burden of a new nomenclature. Another objectionable practice, prevailing quite as extensively also, consists in the change of names or the modification of the extent and meaning of old ones, without the addition of new information or of new views. If this practice is not abandoned it will necessarily end in making Natural History a mere matter of nomenclature, instead of fostering its higher philosophical character. Nowhere is this abuse of a useless multiplication of names so keenly felt as in the nomenclature of the fruits of plants, which exhibits neither insight into vegetable morphology nor even accurate observation of the material facts.[61]

May we not return to the methods of such men as Cuvier and Baer, who were never ashamed of expressing their doubts in difficult cases and were always ready to call the attention of other observers to questionable points, instead of covering up the deficiency of their information by high-sounding words!

In this rapid review of the history of Zoology I have omitted several classifications, such as those of Kaup and Van der Hoeven, which might have afforded an opportunity for other remarks, but I have already extended this digression far enough to show how the standards I have proposed in my second chapter may assist us in testing the value of the different kinds of groups generally adopted in our classifications, and this was from the beginning my principal object in this inquiry.[62] The next step should now be to apply these

[61] [Agassiz was entirely justified in this condemnation of the zoological practices of his time.]

[62] In this edition [1859] of the *Essay on Classification* which is intended as an Introduction to the study of Natural History in general, Janus van der Hoeven's text-book deserves more than a passing notice, especially since its translation by Professor William Clark, *Handbook of Zoology* (1856) is likely to be in the hands of every English student of Natural History. The manner in which the characteristics of the minor groups are presented in this work is so admirable, the reference to the proper authorities so full, the evidence of a personal acquaintance with the objects described so general, and the freedom from mere compilation so praiseworthy, that it is not only to be considered as a text-book for beginners, but truly as a compendium of the present state of Zoology that may be useful even to the professional naturalist. Although taking the views of Cuvier respecting the primary divisions of the Animal Kingdom as a guide, the author does not seem to hold them of such importance, or sufficiently defined, to deserve special consideration. He has thus deprived himself in great measure of the opportunity of presenting in a connected manner those broader generalizations respecting the affinities and homologies of the different classes of animals, which, however, constitute the most important progress of modern Zoology and have secured for our science so important a place among the philosophical studies of our age. It

standards also to the minor divisions of the animal kingdom, down to the genera and species, and to do this for every class singly, with special reference to the works of monographers. But this is such an herculean task, that it can only be accomplished by the combined efforts of all naturalists during many years to come.

---

seems to me also that, though not entirely neglected, the history of the fossil remains is not sufficiently prominent, and the manner in which they are frequently presented, without connection with the living types, is particularly unfavorable to a true appreciation of their natural relation to their living representatives. The time has truly come when the whole Animal Kingdom should be represented in its development through all geological periods as fully as the mode of growth of the living is, in our days, connected with their general history.

Respecting the classes [in Van der Hoeven's classification], I believe, for reasons already stated, that the Infusoria ought to be divided off according to their natural affinity, partly among the Algæ, partly among the Worms, and partly among the Bryozoa. The relation of the Rhizopods to the lower Algæ and especially to the Corallines seems to me daily more probable, and I consider the evidence thus adduced of the vegetable character of the Anentera as amounting almost to a demonstration.

In the class of Acalephs, the Ctenophoræ occupy a position inferior to the Discophoræ. It seems to me hardly questionable that they should occupy the highest position in that class. The Sipunculidæ, which I am inclined to refer to the class of worms, are included among the Echinoderms. The ambulacral system, with or without external suckers, constitutes the essential character of the Echinoderms. Sipunculus has none. The distinction of the intestinal worms and the Annulata as two distinct classes, separated by the Rotatoria, seems to me unnatural. The Turbellaria and Suctoria unite the Annulata with the Trematodes and other worms as one class, and the most recent investigations show unquestionably that the Rotatoria are Crustacea. It seems to me also unnatural to separate the insects and spiders as two classes. The Tardigrada and Acarina form the transition to the Podura and Epizoa. The class of Crustacea, though well defined if we add the Rotatoria to it, should be placed below the insects. The general classification of the branch of Mollusks appears the least satisfactory in this work, for while the Tunicata are considered as a distinct class, and the Conchifera as another, including the Brachipods, the class of Mollusca proper includes not only the Pteropoda and Gasteropoda, but also the Cephalopoda. Evidently, the Cephalopoda are brought here into too close connection with the Gasteropoda. A fuller consideration of the fossil Cephalopoda would easily have satisfied the author that these animals constitute for themselves an independent class.

Since the publication of the *Animal Kingdom* of Cuvier, Van der Hoeven's Textbook is the only general work on Zoology in which the class of fishes is presented in a manner indicating a thorough acquaintance with this class of animals. The treatment of the other Classes of Vertebrata is equally deserving of praise.

INDEX

THE JOHN HARVARD LIBRARY

The intent of Waldron Phoenix Belknap, Jr., as expressed in an early will, was for Harvard College to use the income from a permanent trust fund he set up, for "editing and publishing rare, inaccessible, or hitherto unpublished source material of interest in connection with the history, literature, art (including minor and useful art), commerce, customs, and manners or way of life of the Colonial and Federal Periods of the United States . . . In all cases the emphasis shall be on the presentation of the basic material." A later testament broadened this statement, but Mr. Belknap's interests remained constant until his death.

In linking the name of the first benefactor of Harvard College with the purpose of this later, generous-minded believer in American culture the John Harvard Library seeks to emphasize the importance of Mr. Belknap's purpose. The John Harvard Library of the Belknap Press of Harvard University Press exists to make books and documents about the American past more readily available to scholars and the general reader.